Integrated Wide-Bandwidth Current Sensing

Tobias Funk • Bernhard Wicht

Integrated Wide-Bandwidth Current Sensing

 Springer

Tobias Funk
Dornstetten, Germany

Bernhard Wicht
Gehrden, Germany

ISBN 978-3-030-53252-9   ISBN 978-3-030-53250-5   (eBook)
https://doi.org/10.1007/978-3-030-53250-5

This Springer imprint is published by the registered company Springer Nature Switzerland AG
The registered company address is: Gewerbestrasse 11, 6330 Cham, Switzerland

Preface

High-performance current measurement in a compact and cost-effective design is highly relevant for monitoring, controlling, and optimizing the flow of energy in growth areas such as industrial electronics, renewable energies, smart grids, electromobility, and the Internet of Things. Highly accurate and fast current sensing circuits are a key element specifically for improving energy efficiency, power density, and construction volume. This book addresses the development and implementation of current sensors on system and circuit level that can be integrated on a microchip with a high measurement bandwidth.

As a key approach, this book explores the combination of an optimized fully integrated Rogowski coil and a Hall sensor to enable current sensing from DC to the multi-MHz region. The combination of these two sensor principles, which measure the current by means of the magnetic field and which do not require any post-processing on the technology side, makes it possible to implement a cost-effective and low-loss hybrid current sensor IC. A planar Rogowski coil is intended to be placed besides a power line to measure on-chip signal currents. For off-chip current sensing, a helix-shaped coil offers mounting the chip directly on top of a power line or power module and to measure the current flow underneath. In order to extend the sensing range towards DC, dedicated lateral and vertical Hall devices are developed along with a sensing front-end. A hybrid current sensor for on-chip current sensing combines the Hall path and the Rogowski path and provides wide-bandwidth current sensing from DC up to 71 MHz. Large signal measurements in a power electronics module confirm that the proposed sensing concepts are well suitable for applications like real-time motor control with typical slopes in the order of $100 \, \text{A}/\mu\text{s}$.

The objective of this book is to provide a systematic and comprehensive insight into high-bandwidth current sensing techniques. Both theoretical and practical aspects are covered. Design guidelines are derived for planar and helix-shaped Rogowski coil, for a lateral and vertical Hall element as well as for several sensor front-end circuits. The material will be interesting for design engineers in industry as well as researchers who want to learn about and apply integrated current sensing techniques. This book is based on our research at Reutlingen University, Reutlingen, Germany, and at the Institute for Microelectronic Systems at Leibniz University

Hannover, Hannover, Germany. The work was sponsored by the Federal Ministry of Education and Research Germany (Grand No. 13FH016IX4). We are grateful to all team members for the invaluable support as well as the enjoyable collaboration. A special thanks goes to our families, without their love and support this book would not have been possible.

Dornstetten, Germany   Tobias Funk
Gehrden, Germany   Bernhard Wicht
May 2020

Contents

1 Introduction 1
   1.1 Motivation 1
   1.2 Scope and Outline of This Book 4
   References 6

2 Motivation and Fundamentals... 7
   2.1 Applications for Wide-Bandwidth Current Sensing................... 7
      2.1.1 On-Chip Current Sensing for Integrated Power Stages 7
      2.1.2 Off-Chip Current Sensing for Power Electronic Applications 9
   2.2 Types of Current Sensors 10
      2.2.1 Shunt Resistor... 11
      2.2.2 Sense-FET.. 12
      2.2.3 Magnetoresistance 13
      2.2.4 Fluxgate 14
      2.2.5 Hall Sensor 15
      2.2.6 Rogowski Coil 16
   2.3 Summary 17
   References 18

3 Rogowksi Coil Current Sensor 25
   3.1 Operation Principle 25
   3.2 Rogowski Coil for On-Chip Current Sensing......................... 28
      3.2.1 Mutual Inductance 29
      3.2.2 Parasitic Components... 33
      3.2.3 Design Trade-Offs 38
      3.2.4 Verification with 3D Field Simulation 42
      3.2.5 Layout.. 44
   3.3 Rogowski Coil for Off-Chip Current Sensing 48
      3.3.1 Mutual Inductance 49
      3.3.2 Parasitic Components... 51
      3.3.3 Design Trade-Offs 55
   3.4 Summary 58

        Appendix   59
        References   60

4   **Rogowksi Coil Sensor Front-End**   63
    4.1   Open-Loop Sensing   65
          4.1.1   First Integrator Stage   66
          4.1.2   Second Integrator Stage   67
          4.1.3   Low-Frequency Error Cancelation   69
          4.1.4   Transient Current Sensing   75
    4.2   Closed-Loop Sensing   78
          4.2.1   Forward Path $G(s)$ 80
          4.2.2   Feedback Considerations   81
          4.2.3   AC Signal Compensation   84
          4.2.4   DC Signal Compensation   86
          4.2.5   Noise Cancelation   87
          4.2.6   Overall System   90
    4.3   Open-Loop vs. Closed-Loop Sensing   90
    4.4   Summary   94
        Appendix   95
        References   98

5   **Hall Current Sensor**   101
    5.1   Operation Principle   101
    5.2   Hall Device for On-Chip Current Sensing   103
    5.3   Hall Device for Off-Chip Current Sensing   111
    5.4   Summary   113
        Appendix   114
        References   114

6   **Hall Current Sensor Front-End**   117
    6.1   Signal Modulation for On-Chip Current Sensing   117
    6.2   Signal Modulation for Off-Chip Current Sensing   118
    6.3   Capacitively Coupled Sensor Front-End   120
    6.4   Summary   123
        Appendix   123
        References   124

7   **Hybrid Current Sensor**   125
        References   127

8   **Conclusion**   129

Index   133

Acronyms

List of Abbreviations

CMFF	Common-mode feedforward
CMRR	Common-mode rejection ratio
CMFB	Common-mode feedback
CMOS	Complementary metal–oxide–semiconductor
DPM	Distributed parameter model
EMI	Electromagnetic interference
GaN	Gallium nitride
GBW	Gain-bandwidth
GMD	Geometric mean distance
HFSS	High-frequency structure simulator
IC	Integrated circuit
IGBT	Insulated-gate bipolar transistor
LPM	Lumped parameter model
MR	Magnetoresistive resistor
PCB	Printed circuit board
PEEC	Partial element equivalent circuit
PMIC	Power management IC
PSD	Power spectral density
SiC	Silicon carbide
TCAD	Technology computer-aided design

Symbols

ϵ_{ox}	$\frac{\text{F}}{\text{m}}$	Permittivity of silicon dioxide
μ_{o}	$\frac{\text{Vs}}{\text{Am}}$	Magnetic permeability of free space
μ_{Nwell}	$\frac{\text{cm}^2}{\text{Vs}}$	Carrier mobility of a N-well

ϕ_1		Clock phase 1
ϕ_2		Clock phase 2
ω_0	$\frac{1}{s}$	Natural frequency of a Rogowski coil
ω_{A1}	$\frac{1}{s}$	Bandwidth of amplifier A_1
a	m	Distance between power line and inner segment of Rogowski coil
A	m^2	Area
A_0	m	DC gain of an amplifier
b	m	Distance between power line and outer segment of Rogowski coil
$\vec{B}$	T	Magnetic field
BW	Hz	Bandwidth
BW_{Rog}	Hz	Bandwidth of a Rogowski coil
c	m	Distance for geometric mean distance calculation
C	F	Capacitance
C_0	F	Parasitic capacitance of Rogowski coil
C_A	F	Parasitic capacitance between terminal A of a Rogowski coil and the power line
C_B	F	Parasitic capacitance between terminal B of a Rogowski coil and the power line
C_{hp}	F	Capacitance of a high-pass filter
C_{int}	F	Integration capacitor
C_{lp}	F	Capacitance of a low-pass filter
C_{np}	F	Junction capacitance between N-well and substrate
$C_{overlap}$	F	Parasitic overlap capacitance of Rogowski coil
C_{sub}	F	Parasitic substrate capacitance of Rogowski coil
d	m	Auxiliary variable for the calculation of the geometry factor of a Hall device
D		Damping coefficient
$d_{x,y}$	m	Distance between segment x and segment y of a Rogowski coil
e	m	Auxiliary variable for the calculation of the geometry factor of a Hall device
f_0	Hz	Dominant pole of a Rogowski coil
f_{az}	Hz	Auto-zeroing frequency
f_c	Hz	Corner frequency between $1/f$ noise and thermal noise
f_{ch}	Hz	Copping frequency
F_{el}	N	Electric field strength
f_{Hall}	Hz	Bandwidth of the Hall path
f_{hp}	Hz	Cut-off frequency of a high-pass filter
F_L	N	Lorenz force
f_{lp}	Hz	Cut-off frequency of a low-pass filter
f_{meas}	m	Frequency of the current signal to be measured
f_{p1}	Hz	Dominant pole integrator 1 for open-loop current sensing

f_{p2}	Hz	Dominant pole integrator 2 for open-loop current sensing
$f_{p2,2nd}$	Hz	Second pole of integrator 2 for open-loop current sensing
$f_{Rog,max}$	Hz	Upper limit of the bandwidth of the Rogowski path
$f_{Rog,min}$	Hz	Lower limit of the bandwidth of the Rogowski path
f_{spin}	Hz	Spinning frequency of current spinning for on-chip current sensing
f_{z1}	Hz	Zero of integrator 1 for open-loop current sensing
g	m	Geometric mean distance (GMD)
G		Geometric correction factor of a Hall sensor
g_{inner}	m	GMD between power line and inner segment of Rogowski coil
g_m	$\frac{A}{V}$	Transconductance of an amplifier
g_{outer}	m	GMD between power line and outer segment of Rogowski coil
h_{coil}	m	Height of the helix-shaped Rogowski coil
$I_{bias,A}$	A	Bias current of an amplifier
$I_{bias,H}$	A	Bias current of a Hall sensor
I_{coil}	A	Current in a Rogowski coil
I_{comp}	A	Compensation current for closed-loop current sensing
I_{ext}	A	Excitation current of a fluxgate current sensor
I_{meas}	A	Signal current to be measured
I_{Rhp}	A	Current through the pseudo-resistance in a high-pass filter
L_o	H	Parasitic inductance of a Rogowski coil
l_{coil}	m	Length of a Rogowski coil
$l_{coil,N}$	m	Length of winding N of a Rogowski coil
$l_{coil,tot}$	m	Total length of all segments of a Rogowski coil
L_{Hall}	m	Length of a Hall device
L_S	H	Self-inductance of a Rogowski coil
L_{shunt}	H	Parasitic self-inductance of a shunt resistor
$l_{winding}$	m	Length of all windings of the Rogowski coil
M	H	Mutual inductance
M_-	H	Sum of all negative mutual inductances between the windings of the coil
M_+	H	Sum of all positive mutual inductances between the windings of the coil
M_p	H	Total mutual inductance of a planar Rogowski coil
$M_{p,max}$	H	Maximal mutual inductance of a planar Rogowski coil
$M_{p,N}$	H	Mutual inductance between power line and winding N
M_v	H	Mutual inductance of a helix-shaped Rogowski coil
$M_{x,y}$	H	Mutual inductance between segment x and segment y of the coil

N		Number of windings of Rogowski coil
N_{dop}	cm^{-3}	Doping density of a N-WELL
N_{max}		Maximal possible number of windings of a Rogowski coil
P_{diss}	W	Power dissipation
P_{dop}	cm^{-3}	Doping density of P-substrate
q	C	Electric charge
q_n	C	Negative electric charge
R	$\frac{V}{A}$	Transfer function of a Rogowski coil
R	Ω	Resistance
R_o	Ω	Parasitic resistance of a Rogowski coil
$R_{o,N}$	Ω	Resistance of one winding of the helix-shaped Rogowski coil
$R_{coil,s}$	Ω	Additional series resistance of a Rogowski coil
R_D	Ω	Damping resistor of resistance of a Rogowski coil
R_H	$\frac{1}{C \cdot m^3}$	Hall coefficient
r_H		Hall factor
R_{Hall}	Ω	Resistance of a Hall device
R_{hp}	Ω	Resistance of a high-pass filter
R_{lp}	Ω	Resistance of a low-pass filter
r_{out}	Ω	Output resistance of an amplifier
R_S	Ω	Sheet resistance
R_{shunt}	Ω	Shunt resistance
R_{via}	Ω	Resistance of the via stack of the helix-shaped Rogowski coil
S_A	$\frac{V}{T}$	Sensitivity of a Hall device
$S_{A,Imeas}$	$\frac{V}{A}$	Sensitivity of a Hall device related to the current I_{meas}
SNR		Signal-to-noise ratio
$S_{out,H}$	$\frac{V}{A}$	Sensitivity of the Hall path
$S_{out,hyb}$	$\frac{V}{A}$	Sensitivity of the hybrid current sensor
$S_{out,R}$	$\frac{V}{A}$	Sensitivity of the Rogowski path
S_{Rog}	$\frac{V}{A}$	Sensitivity of a Rogowski coil
$s_{winding}$	m	Space between the windings of a Rogowski coil
t_{chip}	m	Thickness of the chip
t_{Hall}	m	Thickness of the Hall device
t_{ox}	m	Thickness of silicon dioxide
$t_{winding}$	m	Thickness of the winding of the Rogowski coil
v	$\frac{m}{s}$	Drift velocity of an electric charge
V_o	V	Induced voltage in a Rogowski coil
V_{CM}	V	Common-mode voltage
V_{coil}	V	Output voltage of a Rogowski coil
V_{comp}	V	Compensation voltage for closed-loop current sensing
V_{Hall}	V	Output voltage of a Hall device

$V_{\text{Hall,off}}$	V	Offset voltage of a Hall device
$V_{\text{Hall,sig}}$	V	Hall voltage caused by a magnetic field
V_{in}	V	Input voltage of a voltage converter
V_{OS}	V	Offset voltage of an amplifier
V_{out}	V	Output voltage of a voltage converter
$V_{\text{out,H}}$	V	Output voltage of a Hall path
$V_{\text{out,hp}}$	V	Output voltage of a high-pass filter
$V_{\text{out,Int1}}$	V	Output voltage of integrator 1
$V_{\text{out,lp}}$	V	Output voltage of a low-pass filter
$V_{\text{out,R}}$	V	Output voltage of a Rogowski path
$V_{\text{out,R,noise}}$	V	Effective noise voltage of the Rogowski path
$V_{\text{out,R,OS}}$	V	Offset voltage at the output of a Rogowski path
V_{sense}	V	Voltage of the sense coils of a fluxgate
V_{shunt}	V	Voltage drop across a shunt resistor
V_{sw}	V	Voltage of the switching node in a buck converter
W	m	Width of a transistor
w_{coil}	m	Width of the Rogowski coil
W_{Hall}	m	Width of a Hall device
$w_{\text{powerline}}$	m	Width of power line
w_{winding}	m	Width of winding of the Rogowski coil
Y_{o}	$\dfrac{\text{A}}{\text{V}}$	Admittance of a Rogowski coil

Chapter 1
Introduction

1.1 Motivation

The continuous trend towards higher integration of power electronic circuits opens up many new applications. Several functions are combined in a compact power module or integrated on a single chip to become smaller, save cost, and to increase the performance. Considerable changes are about to happen in the control and monitoring of energy in several interconnected areas like power generation, e-mobility and automotive, smart grid, automation, Internet of Things, and others, as indicated in Fig. 1.1.

In power generation or smart grid applications, for example, energy must be converted or distributed as required. In smart home applications or in electric vehicles, voltage conversion from a high supply level to a lower voltage level is needed. Thus, the sensing of electrical current plays an important role for these power electronic applications. It enables the demand driven power supply or the monitoring of the functionality of the system. Due to the continuing trend towards higher integration densities and higher supply levels, the requirements for low-loss current sensing with high bandwidth on IC level are simultaneously increasing.

Voltage converters, in particular, are becoming more and more compact and the number of passive components is decreasing. This can be achieved by increasing the switching frequency, for example, which makes the dominant passive components smaller and allows full integration [1, 2]. For the control of a voltage converter it is essential to monitor the current with a bandwidth, which is a multiple of the switching frequency. Therefore, a fully integrated wide-bandwidth current sensor is required for the continuous development of such voltage converters.

By using power switches like Insulated-gate bipolar transistors (IGBTs) or new power semiconductors such as Silicon carbide (SiC) or Gallium nitride (GaN) the power supply levels can be increased to higher voltages [3, 4]. This reduces load currents and enables a more efficient energy transfer, as it is also done between

© The Editor(s) (if applicable) and The Author(s), under exclusive license to
Springer Nature Switzerland AG 2020
T. Funk, B. Wicht, *Integrated Wide-Bandwidth Current Sensing*,
https://doi.org/10.1007/978-3-030-53250-5_1

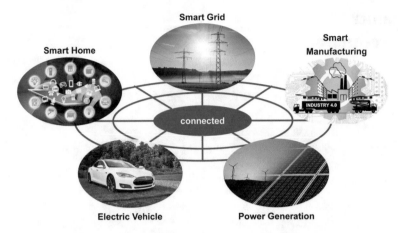

Fig. 1.1 Interconnection and development of different power electronics applications

the energy generators and the grid [5]. In addition, due to their better switching characteristics, these power switches also enable to increase the switching frequency of high-voltage converters, which results in a reduction in size and cost of passive components. In applications with these power devices, it is important to detect the resulting fast changing current slopes directly in the power module for control and fault monitoring. Error detection is necessary, as otherwise the electronics could be destroyed by incorrect switching of the power device. Thus, a compact wide-bandwidth current sensor is essential for the continuous development of such power electronics applications.

The use of silicon technologies enables a cost-effective implementation of sensors. Modern silicon based sensors combine the sensing element and the front-end circuits on a single chip, called smart sensors [6]. Therefore, the output signal of the sensor element, often in the microvolt range, can be measured more accurately. Furthermore, the parasitics between the sensing element and the front-end are minimal, resulting in lower losses and better performance. At the same time, the use of silicon creates challenges in the design of the front-end, as low-frequency error sources such as $1/f$ noise and offset must be taken into account. These low-frequency errors can basically be addressed by two dynamic error cancelation techniques: sample-and-correct technique and modulate-and-filter technique [7]. Auto-zeroing, for example, is a sample-and-correct technique, in which the input of an amplifier is periodically shorted and the resulting error at the output is stored on a capacitor. Subsequently the stored error voltage of the amplifier will be subtracted from the output signal. Chopping is a very popular modulate-and-filter technique to reduce low-frequency error signals. The output signal of the sensing element is modulated to a rectangular signal, amplified and then demodulated again.

Integrated current sensing can be implemented with several principles. For energy-efficient and wide-bandwidth current sensing, a shunt-based sensing needs a precise sense resistor with low impedance in addition to low parasitic inductance. Current sensors like Hall, fluxgate, and Magnetoresistive resistor (MR) detect the

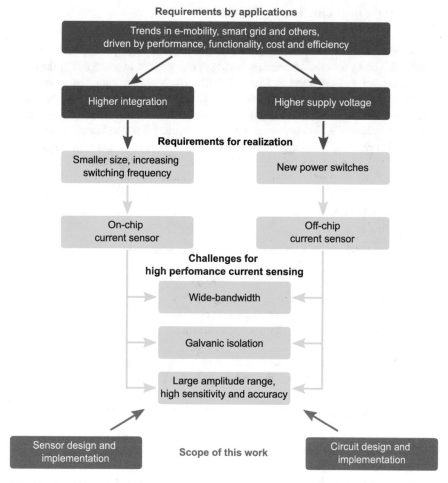

Fig. 1.2 Summary of the scope of this work

current by means of its magnetic field $\vec{B}$. They provide galvanic isolation, required for many power applications, but have limited bandwidth at frequencies below 500 kHz and often require extensive post-processing. Current sensing based on a Rogowski coil is well established for discrete current sensing and was even verified in a planar PCB-based application with very high bandwidth. It provides cost-efficient integration without post-processing, while offering galvanic isolation and superior bandwidth. However, since no DC currents can be measured, a hybrid current sensor is required to extend the limited low-frequency range of a Rogowski coil towards DC signal currents. Therefore, the combination of an integrated Rogowski coil and a Hall sensor can provide a cost-effective current sensor that reaches the wide-bandwidth requirements and can follow the continuing trend towards high integration.

1.2   Scope and Outline of This Book

The scope of this book, based on the trend towards high-performance current sensing driven by the control and monitoring of energy flow in multiple applications, is summarized and depicted in Fig. 1.2. Current sensors are required to measure and replicate fast changing current profiles. This work covers galvanically isolated current sensors for on-chip and off-chip current sensing in standard silicon technologies. By combining a Hall sensor and a fully integrated Rogowski coil in one chip, DC signal currents and fast current changes can be measured with a wide bandwidth. The largest challenges are the required bandwidth, the small sensor signals of both sensor principles, the wide dynamic range of a Rogowski coil, and the reduction of low-frequency errors of the sensor front-end in a high-voltage CMOS technology. These challenges are addressed in this work in the following way:

1. A combination of two sensor principles is investigated. The Hall sensor measures DC and low-frequency signal currents and a Rogowski coil measures high-frequency signal currents.
2. Layout and geometry of the sensing elements are investigated. The sensitivity of a Hall device depends strongly on the geometry. For the design of a Rogowski coil, parasitic components are taken into account in order to achieve a high bandwidth with high sensitivity at the same time.
3. Multi-stage amplification of sensor signals over a wide frequency range is investigated. For the Rogowski path, a new compensating sensor front-end is explored in order to increase the sensitivity and to extend the linear sensing range at high frequencies.
4. Low-frequency error cancelation techniques are applied to reduce noise and offset.

The outline of this book is as follows. Chapter 2 describes the motivation for this work, and the requirements for wide-bandwidth current sensing. Applications for on-chip and off-chip current sensing are reviewed in Sect. 2.1. The requirements for on-chip current sensing, especially for fully integrated fast switching voltage converters operating in the multi-MHz region, are discussed in Sect. 2.1.1 and the bandwidth requirements are derived. Power switches such as IGBT or new semiconductors such as GaN and SiC are used in high-voltage and high-power applications. The resulting bandwidth requirements for off-chip current sensing to replicate the fast changing current slopes are introduced in Sect. 2.1.2. Different state-of-the-art current sensors principles are reviewed in Sect. 2.2, pointing out the advantages of individual sensing principle in terms of bandwidth, sensitivity, and cost.

Chapter 3 elaborates the sensing of fast changing signal currents by means of a fully integrated Rogowski coil. The fundamentals of a Rogowski coil and the dependencies for design in terms of bandwidth and sensitivity are derived in Sect. 3.1. For the optimized implementation of a planar Rogowski coil for on-chip current sensing, a model is introduced in Sect. 3.2. The conventional equivalent circuit diagram is extended by coupling capacities, which are relevant for the fully integrated implementation of a Rogowski coil. The design trade-offs between bandwidth and sensitivity are discussed and the layout of the coil is optimized with respect to capacitive coupling. Various planar Rogowski coils are examined with 3D-field simulations by Ansys HFSS (High Frequency Structure Simulator). An optimized helix-shaped Rogowski coil for off-chip current sensing is introduced and the capability of back-grinded chips is evaluated in Sect. 3.3. This allows to mount the current sensor directly on top of the power line and measure the current flowing underneath.

In Chap. 4, different sensor front-ends for signal processing and a comparison between auto-zeroing and chopping for noise reduction are proposed. An open-loop two-stage integrator circuit, introduced in Sect. 4.1, enables wide-bandwidth current sensing and chopping for noise and offset reduction. In contrast to conventional chopping, the chopper frequency can be selected lower than the signal bandwidth. A compensated sensor front-end for wide-bandwidth closed-loop current sensing improves offset, sensitivity, and sensing range at high frequencies and is proposed in Sect. 4.2. Auto-zeroing and chopping are used for noise reduction. Section 4.3 shows the comparison between the introduced sensor front-ends and Sect. 4.4 the comparison of the implemented on-chip and off-chip current sensing with state-of-the-art current sensors.

In Chap. 5, DC and low-frequency current sensing with a Hall sensor is described, since this frequency range cannot be covered with a Rogowski coil. The fundamentals of a Hall device and the dependencies for the design in terms of sensitivity are derived in Sect. 5.1. Different symmetrical layouts of planar Hall devices for on-chip current sensing are reviewed in Sect. 5.2. For an optimal design of the Hall device, parameters of the used technology are extracted from measurement results and used in Silvaco TCAD (Technology Computer-Aided Design) simulations. For off-chip current sensing, the implementation of a vertical Hall device is described in Sect. 5.3.

A capacitively coupled sensor front-end is introduced in Chap. 6. The cross-shaped planar Hall device in Sect. 6.1 allows current spinning for offset reduction of the Hall device itself and noise and offset reduction of the following sensor front-end. Chopping is implemented for off-chip current sensing with a vertical Hall device in Sect. 6.2.

The combination of the introduced current sensors to a hybrid current sensor is described in Chap. 7. DC and low-frequency signal currents are measured with the Hall path, while fast changing signal currents are covered by the Rogowski path. These two sensor principles are connected by a passive filter and enables wide-bandwidth current sensing.

Chapter 8 summarizes and concludes the results of this work.

References

1. Ghahary, A. (2004). *Fully integrated DC-DC converters*. https://www.powerelectronics.com/content/fully-integrated-dc-dc-converters.
2. Wittmann, J., Funk, T., Rosahl, T., & Wicht, B. (2018). A 48-V wide-V_{in} 9–25-MHz resonant DC–DC converter. In *IEEE Journal of Solid-State Circuits, 53*(7), 1936–1944. ISSN:0018-9200. https://doi.org/10.1109/JSSC.2018.2827953.
3. Singh, R. (n.d.). *Silicon carbide switches in emerging applications*. https://www.psma.com/sites/default/files/uploads/tech-forums-semiconductor/presentations/16-silicon-carbide-switches-emerging-applications.pdf.
4. Efficient Power Conversion. (2018). *Where is GaN going?* https://epc-co.com/epc/GalliumNitride/whereisgangoing.aspx.
5. Beta. (2017). *Transmitting electricity at high voltages*. http://www.betaengineering.com/high-voltage-industry-blog/transmitting-electricity-at-high-voltages.
6. Makinwa, K. (2014). Smart sensor design. In *Smart sensor systems*. Hoboken: John Wiley & Sons, Ltd. (Chap. 1, pp. 1–16). ISBN:9781118701508. https://doi.org/10.1002/9781118701508.ch1. eprint: https://onlinelibrary.wiley.com/doi/pdf/10.1002/9781118701508.ch1. https://onlinelibrary.wiley.com/doi/abs/10.1002/9781118701508.ch1.
7. Wu, R., Huijsing, J. H., & Makinwa, K. A. A. (2013). *Precision instrumentation amplifiers and read-out integrated circuits*. Berlin: Springer. ISBN:9781461437307. https://doi.org/10.1007/978-1-4614-3731-4.

Chapter 2
Motivation and Fundamentals

2.1 Applications for Wide-Bandwidth Current Sensing

In this book, two types of current sensing are considered. Some applications require current sensing directly on IC level to measure signal currents inside the chip. This saves costs by reducing the number of pins of a chip and the number of external components, and it increases reliability and accuracy by reducing the effect of parasitics. Applications with high power and discrete power stages require an off-chip current sensor that measures signal currents without loss and provides the necessary galvanic isolation to high-voltage levels. In this section, an overview of applications for on-chip and off-chip current sensing is given and the bandwidth requirements are derived.

2.1.1 On-Chip Current Sensing for Integrated Power Stages

Applications with integrated power stages such as class-D audio amplifiers or voltage converters require an on-chip current sensor to measure internal signal currents for control and fault detection.

Class-D amplifiers are used to drive loudspeakers with high-power efficiency. A pulsed signal is generated in the power stage of the class-D amplifier, which is then filtered with a passive filter before being transmitted to the speaker [1]. It is important to monitor the current flow through the speakers, as otherwise excessive heat may be generated in the voice coil, causing permanent damage to the speakers. The clock frequency of the pulsed signal is typically in the range of several hundred kHz, which requires current sensing with a high bandwidth of $>10\,\text{MHz}$ [2]. In order to reduce these high-bandwidth requirements, the current can be measured after the filter. However, this requires additional pins, which prevents cost-effective

© The Editor(s) (if applicable) and The Author(s), under exclusive license to
Springer Nature Switzerland AG 2020
T. Funk, B. Wicht, *Integrated Wide-Bandwidth Current Sensing*,
https://doi.org/10.1007/978-3-030-53250-5_2

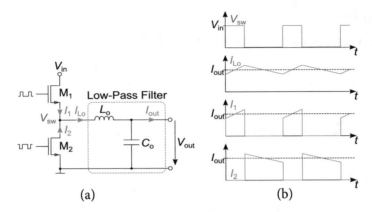

Fig. 2.1 Buck converter: (**a**) schematic and (**b**) simplified waveforms

implementation. Alternatively, the current through the speakers can be measured directly on IC level. In this case, the current in the power stage is mirrored to a sense path. However, the maximum bandwidth is not sufficient because it is limited to less than ≈5 MHz for amplifiers with high supply voltages, as will be discussed in Sect. 2.2.2.

The trend towards higher integration density can be observed in particular in Power management ICs (PMICs). Due to new circuit concepts of switching voltage converters and higher switching frequencies in the multi-MHz region, external components can be greatly reduced or even saved.

Inductive voltage converters provide voltage conversion by periodically energizing an inductance. A buck converter, for example, consists of a power stage connected to V_{in} and a low-pass filter with an inductor and a capacitor, see Fig. 2.1a. The power stage (M_1, M_2) generates a pulsed signal V_{sw} with the amplitude of V_{in} (see Fig. 2.1b), which is passed through a low-pass filter. This results in a nearly constant output voltage V_{out}, which can be regulated by the duty cycle of the switching voltage V_{sw}.

Many control methods for inductive DCDC converters measure the current I_{Lo} through the inductance. This is often done by measuring the voltage drop across the inductance originating from its parasitic series resistance. As a major disadvantage, the frequency response of the parasitic series resistance must be compensated by additional passives [3, 4], as will also be investigated in Sect. 2.2.1 for current sensing with a shunt. Furthermore, additional bond pads are required to connect the external inductance to the integrated circuit and the current sensor is more sensitive to interference signals. Alternatively, the current through the inductance can be measured directly in the power stage, which eliminates the need for an additional compensation network. In this case, the currente thought the switching network (I_1 and / or I_2) is mirrored to a sense path. However, this has a limited bandwidth of less than ≈5 MHz and is only possible with great effort for converters with higher input voltages [5, 6].

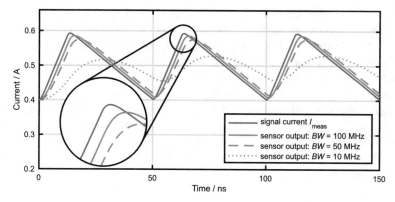

Fig. 2.2 Transient response of modeled current sensors with different bandwidths BW for a voltage converter with 200 mA signal current ripple and 20 MHz switching frequency

Advanced fast switching converters for high input voltages up to 100 V reach switching frequencies up to 10 MHz [7–10]. In resonant converters, the frequency can be even higher up to 25 MHz [11, 12]. Figure 2.2 shows the current $I_{meas} = I_{Lo}$ through the inductor of a buck converter and the response of a current sensor, which are ideally modeled, except for different bandwidths BW. The buck converter operates at a switching frequency of 20 MHz with an output current of 0.5 A and a ripple of 0.2 A. Such converters require a current sensor with a bandwidth BW, which is a multiple of the switching frequency of the converter. Only a current sensor with a bandwidth of 50 MHz or even 100 MHz can reproduce the signal current I_{meas} with sufficient accuracy and low delay, as shown in Fig. 2.2.

In conclusion, applications with integrated power stages such as class-D amplifiers or voltage converters require high-bandwidth current sensing. For inductive voltage converters, for example, which operate with a high switching frequency of 20 MHz, a current sensing bandwidth from DC up to more than 50 MHz is required in order to reproduce the current through the inductor with sufficient accuracy. By measuring the current by means of the magnetic field $\vec{B}$ the capacitive coupling between the switching node and the sensing element can be suppressed. In addition, the number of pins can be reduced by a fully integrated on-chip current sensor.

2.1.2 Off-Chip Current Sensing for Power Electronic Applications

Several electronic applications tend towards higher voltages and higher power. The focus is on the conversion of electrical energy or the operation of large loads such as voltage converters for solar cells or motor drivers. In order to fulfill the high-power requirements, power switches such as Insulated-gate bipolar transistor (IGBT)s or, more recently, wide-bandgap semiconductors such as Gallium nitride

(GaN) and Silicon carbide (SiC) are used [13, 14]. GaN power switches can handle voltages up to 600 V, while IGBT and SiC devices can even handle voltages $>1\,\text{kV}$. With these higher operating voltages, galvanic isolation is essential and due to the simultaneously improved switching characteristics of these power devices, fast switching slopes of several kA/µs have to be measured.

Voltage conversion in high-power applications operates basically like integrated voltage conversion, introduced in Sect. 2.1.1 (Fig. 2.1a). However, IGBT, GaN, and SiC devices provide high-voltage and high-current capabilities. At the same time, lower parasitic capacitances reduce switching losses, while faster switching transitions can be achieved. Therefore, wide-bandwidth current sensing is required for the faster current transitions and the increasing switching frequencies of the converter.

In large electrical machines, voltages of typically several hundred volts are commutated with an IGBT power module [15]. Such power modules require a current sensor directly at the switch for control purposes in order to measure the current through the device without distortion by parasitic effects.

Other applications such as telecommunications, medical applications, wireless power supply, or e-mobility do not require this very high dielectric strength, but these applications also benefit from faster and more efficient switching.

In power applications, the detection of faults is mandatory, since the consequences of destruction are larger due to the higher voltages and currents. Furthermore, there is a trade-off between the switching speed and the resulting Electromagnetic interference (EMI) in these applications. For this reason, the current slope ($\mathrm{d}I/\mathrm{d}t$) of the switch must be measured and regulated depending on the application [15, 16]. Some power devices have a dedicated sense contact for this purpose. This allows determination of the derivation $\mathrm{d}I/\mathrm{d}t$ based on the voltage change across the parasitic inductance of the package. However, there are several challenges to this approach. For instance, the measurement system needs to be tuned to the amount of parasitic inductance, which depends on the device package [17]. Hence, the device cannot be replaced arbitrarily.

In conclusion, in power switches current slopes of several kA/µs can occur, which are to be measured for fault detection and $\mathrm{d}I/\mathrm{d}t$ regulation. Figure 2.3 shows exemplary the signal current I_{meas}, with an amplitude of 100 A and a rise of 3 kA/µs, and the response of ideally modeled current sensor with different bandwidths BW. Only by a current sensor with a bandwidth above 10 MHz or even 20 MHz it is possible to reproduce the signal current I_{meas} with sufficient accuracy and low delay.

2.2 Types of Current Sensors

Current sensors for on-chip and off-chip current sensing can be realized with several principles and can be divided into galvanically isolated and nonisolated sensors.

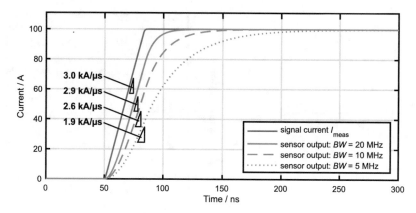

Fig. 2.3 Step response of current sensors with different bandwidths BW for a 100 A signal current pulse with a 3 kA/μs rise time

In this section, a study on state-of-the-art current sensors is carried out and the applicability for fully integrated wide-bandwidth current sensing is investigated.

2.2.1 Shunt Resistor

A classical and widely used principle for on-chip and off-chip current sensing is the sensing of the voltage drop across a shunt resistor [18–22]. Unfortunately, the resulting voltage drop causes losses and the parasitic properties limit the maximum bandwidth of the signal. To compensate for the influence of the parasitic series inductance L_{shunt} and to increase the bandwidth of the shunt, a compensation network can be added [22], as shown in Fig. 2.4a. In this case, the time constant of the shunt R_{shunt} and L_{shunt} must correspond to the time constant of the compensation network:

$$\frac{L_{shunt}}{R_{shunt}} = R_{lp}C_{lp} \tag{2.1}$$

Figure 2.4b shows the measurement results of a conventionally available discrete shunt resistor of 33 mΩ and a compensation network with $R_{lp} = 910\,\Omega$ and $C_{lp} = 1$ nF. At high signal frequencies f_{meas} the measured voltage V_{shunt} increases significantly due to the increasing impedance of L_{shunt}, whereas the measured output voltage $V_{out,lp}$ of the compensation network remains constant over the total frequency range. Thus, the signal current I_{meas} can be correctly replicated even at high signal frequencies f_{meas}.

With this compensation network, the bandwidth of the shunt is significantly extended, but losses at high frequencies are still increasing. Therefore, current sensing with a shunt resistor in the power line is not preferred for power electronics

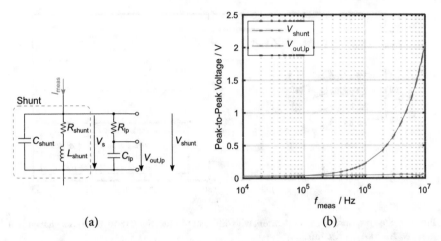

(a)   (b)

Fig. 2.4 Shunt current sensor: (**a**) equivalent circuit diagram with compensation network and (**b**) measurement results for shunt resistor $R_{\text{shunt}} = 33\,\text{m}\Omega$ and a signal current I_{meas} with an amplitude of 1 A

applications with large signal currents and high frequencies. Furthermore, an IC level implementation of the compensation network is not usual due to the large tolerances of the absolute values of R_{lp} and C_{lp} in fabrication. A shunt is particularly suitable for the current sensing of low-frequency currents ($<100\,\text{Hz}$) with a high resolution of $200\,\mu\text{A}$ [20]. Unfortunately, such a high resolution requires a temperature monitoring of the shunt at the same time, as the shunt heats up due to the resulting power loss.

Alternatively to a shunt resistor, the power line can be used, whereby no additional series inductance occurs [21, 22]. In this case, high-precision circuits allow accurate measurement of signal currents from a few microamp up to several amps, but they have a limited bandwidth of $<25\,\text{kHz}$ due to the signal processing and also require temperature monitoring of the power line.

2.2.2   Sense-FET

For fully integrated on-chip current sensing, the accurate and lossless sense-FET technique is widely used. In switching applications, a sense-FET M_s is connected in parallel with the power-FET M_1. The width W_{MS} of M_s is a small fraction of the width W_{M1} which results in sensing a small portion of the signal current I_{meas}.

Figure 2.5 shows the principle of high-side sense-FET current sensing circuit. The ratio between the power-FET and the sense-FET is in the range of several hundred to several thousand [6, 23–25]. The amplifier A_1 is necessary for forcing the drain-source voltage to be equal and therefore to improve the accuracy of the current mirror. In state-of-the-art sense-FET implementations, a high bandwidth

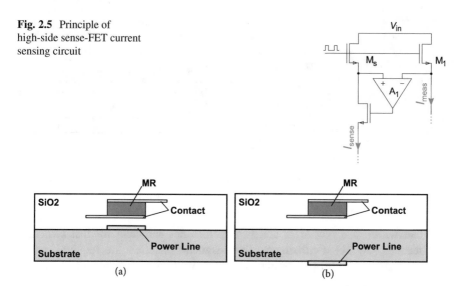

Fig. 2.5 Principle of high-side sense-FET current sensing circuit

Fig. 2.6 Magnetoresistive current sensor for (**a**) on-chip current sensing and (**b**) off-chip current sensing

>10 MHz is reached for low voltage applications (<5 V) [25, 26], while for high-voltage applications up to 40 V the bandwidth is limited to below 7 MHz [5, 6]. This principle is restricted to on-chip current sensing and may not be used for off-chip current sensing.

2.2.3 Magnetoresistance

Magnetoresistive resistors (MR) are suitable for galvanically isolated and lossless current sensing. Due to the magnetoresistive effect, the change of the resistance value depends on the external magnetic field $\vec{B}$. Several of these resistors are typically connected as a Wheatstone bridge to increase the sensitivity and to be insensitive to interference fields [27–30]. MRs are particularly characterized by their high sensitivity and accuracy. This even allows to measure quiescent currents of integrated circuits to verify the presence of manufacturing faults (IDDQ test) [27].

Magnetoresistive resistors (MRs) are suitable for on-chip and off-chip current sensing. Figure 2.6a shows the simplified cross-section for on-chip current sensing [27], whereby the MR is placed above the power line and provides the galvanic isolation required for switching applications. For off-chip current sensing, the chip with the integrated MR can be placed on top of the power line and measures the current flowing underneath [29, 30], as indicated in Fig. 2.6b.

The MR has an inherent bandwidth in order of GHz, which would fulfill all requirements outlined in Sect. 2.1 [28]. However, signal processing limits

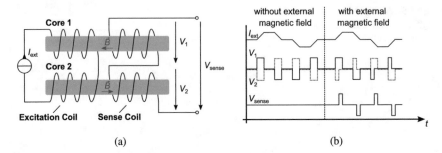

Fig. 2.7 Fluxgate current sensor with **(a)** excitation coil and sense coil and **(b)** simplified waveforms

the bandwidth to a few MHz and is therefore not sufficient for fast switching applications [28–31]. Furthermore, ferromagnetic materials are required, which results in expensive post-processing and thus prevents cost-efficient implementation of the current sensor.

2.2.4 Fluxgate

A fluxgate senses the signal current to be measured by means of the magnetic field and thus provides inherent galvanic isolation [32–34]. It consists of two ferromagnetic cores, which are periodically driven in and out of saturation by excitation coils and by the bias current I_{ext}, as indicated in Fig. 2.7a. Typically, the cores are placed in parallel and are saturated in an antiparallel way by the orientation of the coils. Thus, the magnetic field in one core is amplified by the external magnetic field and reduced in the other core. If no external magnetic field is present, the two induced voltages V_1 and V_2 in the sensing coils canceled out each other. However, in presence of an external magnetic field, one core is saturated sooner than the other, which results in a signal voltage V_{sense} at the sensing coil, as indicated in Fig. 2.7b [32]. The pulse-width of V_{sense} is a measure of the magnetic field to be measured. By demodulating and integrating V_{sense}, the signal current to be measured can be reproduced.

A fluxgate has a high sensitivity and signal currents <1 mA can be measured. However, the bandwidth is limited to <75 kHz by the complex signal processing and the periodic saturation of the cores [34]. Furthermore, it is restricted to off-chip current sensing and requires expensive post-processing due to the need of two ferromagnetic cores.

Fig. 2.8 Operation principle
of a Hall device

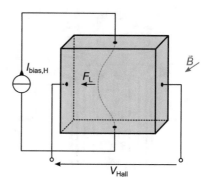

2.2.5 Hall Sensor

A Hall sensor offers the opportunity to determine the current to be measured
by means of the magnetic field without expensive post-processing. The so-called
Hall effect, outlined in Fig. 2.8, generates a measurable voltage V_{Hall}, which is
proportional to the magnetic field $\vec{B}$ and the bias current $I_{\text{bias,H}}$. The current flow
through a conductive material is deflected perpendicular to the magnetic field $\vec{B}$
and to the direction of current $I_{\text{bias,H}}$ due to the Lorenz force F_{L}. This charge shift
creates an electric field against the Lorenz force, which can be measured in form of
the Hall voltage V_{Hall}, as indicated in Fig. 2.8.

In general, any conducting material can be used as a Hall device. However,
semiconductors have a lower charge carrier density, e.g. compared to copper, which
results in a higher sensitivity [35]. In silicon, Hall sensors are typically implemented
in N-wells. This has the further advantage that both the Hall device and the
required sensor front-end can be implemented in one chip [36–40]. Unfortunately,
the sensitivity of the Hall device strongly depends on the conductivity of the active
N-well and thus also on the temperature. This leads to a complex temperature
drift compensation or to a continuous real-time calibration of the sensitivity by
the bias current $I_{\text{bias,H}}$ [39–44]. Furthermore, gradients in the active N-well, e.g.
in the doping or in the temperature, produce an offset voltage, which must also
be compensated by the signal processing. Thus, the bandwidth of a Hall sensor
is typically limited to <500 kHz [38]. Nevertheless, a Hall device is often used
as a current sensor due to its simple design and high flexibility. The sensitivity
can be easily adjusted by the bias current $I_{\text{bias,H}}$, which is particularly suitable for
integration into a hybrid current sensor [45–49]. Furthermore, different designs of
Hall devices offer the possibility for on-chip current sensing [36–40, 50, 51] and
off-chip current sensing [52–57].

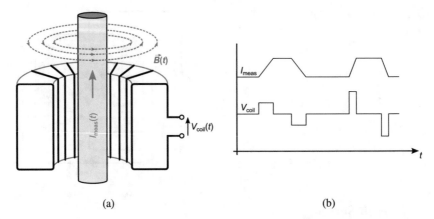

Fig. 2.9 Cross-section of (**a**) a discrete Rogowski coil and (**b**) simplified waveform of the induced voltage

2.2.6 Rogowski Coil

A Rogowski coil is an air coil that encloses the conductor to be measured, as shown in Fig. 2.9a. This is based on the method proposed by Walter Rogowski in 1912 [58]. A changing magnetic field $\vec{B}$ induces the voltage V_{coil} into the coil which is a measure of the signal current I_{meas} to be measured [58–60]. The induced voltage V_{coil} is proportional to the strength and the time change of the magnetic field $\vec{B}$, as indicated in Fig. 2.9b. Thus, the Rogowski coil has a differentiating transfer behavior, whereby only AC signal currents or current changes can be measured.

For the same reason, a Rogowski coil is particularly suitable for current sensing with a high bandwidth up to 100 MHz [61]. At the same time very large currents can be measured, since the Rogowski coil consists of an air coil and thus no ferromagnetic material gets into saturation [62]. The sensitivity and the bandwidth for current sensing are mainly limited by the design of the coil. The cross-sectional area of the coil and the distance to the current conductor to be measured determine the sensitivity, while the resonance frequency of the coil limits the bandwidth for current sensing. In order to replicate the signal current I_{meas}, the differentiating transfer behavior of the coil must be compensated by a sensor front-end with an integrating transfer behavior [63–67]. Due to the high dynamic range (number of decades in the frequency range corresponds to number of decades in the voltage range) and the high bandwidth of the Rogowski coil, high demands are placed on signal processing in this context.

After 2004, the operation principle of a Rogowski coil was successfully transferred to PCB implementations [61, 68–72]. The current conductor to be measured is enclosed by different arrangements of single coils or a planar coil is placed besides a power line and measures currents on the PCB. This enables a small distance between the coils and the current conductor and thus a high sensitivity.

Since only AC signal currents can be measured with a Rogowski coil, but DC current detection is required for most applications, the Rogowski coil is often combined with other sensor principles to form a hybrid current sensor. Thus, for example, the limited bandwidth of a Hall sensor or MR can be extended by a Rogowski coil [45–47, 62, 73, 74].

Similar to the planar PCB level implementation, the realization of an air coil is also possible on IC level. This represents the key approach of this work. With different coil designs, this principle is suitable for on-chip and off-chip current detection. Initial publications demonstrate that the bandwidth of an integrated Hall sensor can be extended [48, 49]. However, the bandwidth in prior art is limited to less than 3 MHz due to a non-optimized coil design.

2.3 Summary

A trend towards high-performance current sensing, driven by control and monitoring of energy flow can be observed in several applications. For on-chip current sensing of a voltage converter, e.g. operating at a switching frequency in the multi-MHz range, a bandwidth from DC up to more than 50 MHz is required to reproduce signal currents with sufficient accuracy. Furthermore, a fully integrated current sensor with a galvanic isolation can suppress the capacitive coupling between the switching node and the sensor element and reduce the number of pins of the chip.

Power electronic applications additionally tend to higher voltages and currents. To satisfy the resulting requirements in terms of switching losses and speed, power switches such as IGBT, SiC, and GaN are used. In order to replicate the signal current with sufficient accuracy, a galvanically isolated off-chip current sensor with a bandwidth from DC up to more than 10 MHz is required to reproduce current slopes up to 3 kA/µs.

Integrated CMOS current sensors are available for on-chip and off-chip current sensing, but have limitations in terms of bandwidth and cost-efficient implementation. Measurements show that a shunt with an adapted compensation network allows current sensing with a high bandwidth (up to 10 MHz). The shunt is suitable for on-chip and off-chip current sensing, but high losses can occur and there is no galvanic isolation. A sense-FET implementation allows only on-chip current sensing and the bandwidth is limited below 7 MHz for high-voltage applications. Current sensors such as MR or fluxgate detect the current by means of its magnetic field $\vec{B}$ and provide the required galvanic isolation. Unfortunately, the bandwidth is limited to a few MHz by signal processing and these sensor principles require a complex post-processing with ferromagnetic materials. A Hall sensor, on the other hand, offers a cost-effective implementation without post-processing for on-chip and off-chip current sensing. Due to the required signal processing, however, the bandwidth is also limited to less than 500 kHz. Figure 2.10 shows the bandwidth limitations of different current sensor principles, along with the desired bandwidth for the current sensing in this book. In comparison to all other methods, the principle of

Fig. 2.10 Bandwidths of
state-of-the-ate current sensor
principles

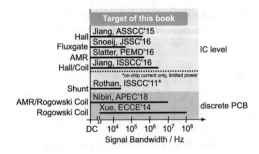

the Rogowski coil allows a cost-efficient current sensing with a high bandwidth (up
to 100 MHz). By different coil designs, on-chip and off-chip current sensing can be
implemented.

Both, the Hall sensor and the Rogowski coil provide a cost-efficient implementation as well as flexibility for on-chip and off-chip current sensing. For this reason,
this work focuses on these two sensor principles. These principles can be designed
in such a way that the limited bandwidth of one principle can be extended by the
other principle. Finally, the two sensors can be connected to form a hybrid current
sensor, which enables current sensing from DC up to several tens of MHz.

References

1. Karaca, T., & Deutschmann, B. (2015). Electromagnetic Evaluation of Class-D Switching
   Schemes. In *Proceedings of the 11th Conference on Ph.D. Research in Microelectronics and
   Electronics (PRIME)* (pp. 113–116). https://doi.org/10.1109/PRIME.2015.7251347.
2. Prakash, A. (2017). *Current sense amplifiers in class-D audio subsystems*. Dallas: Texas
   Instruments. http://www.ti.com/lit/an/slya031/slya031.pdf.
3. Herzer, S., Kulkarni, S., Jankowski, M., Neidhardt, J., & Wicht, B. (2009). Capacitive-coupled
   current sensing and auto-ranging slope compensation for current mode SMPS with wide supply
   and frequency range. In *2009 Proceedings of the ESSCIRC* (pp. 140–143). https://doi.org/10.
   1109/ESSCIRC.2009.5326034.
4. Ahsanuzzaman, S. M., & Prodić, A. (2015). An on-chip integrated auto-tuned hybrid current-
   sensor for high-frequency low-power DC-DC converters. In *Proceedings of the IEEE Applied
   Power Electronics Conference and Exposition (APEC)* (pp. 445–450). https://doi.org/10.1109/
   APEC.2015.7104388.
5. Dake, T., & Ozalevli, E. (2008). A precision high-voltage current sensing circuit. *IEEE
   Transactions on Circuits and Systems I: Regular Papers, 55*(5), 1197–1202. ISSN: 1549-8328.
   https://doi.org/10.1109/TCSI.2008.916452.
6. Renz, P., Lamprecht, P., Teufel, D., & Wicht, B. (2016). A 40V current sensing circuit with
   fast on/off transition for high-voltage power management. In *Proceedings of the IEEE 59th
   International Midwest Symposium on Circuits and Systems (MWSCAS)* (pp. 1–4). https://doi.
   org/10.1109/MWSCAS.2016.7870011.
7. Xue, J., & Lee, H. (2016). A 2MHz 12-to-100V 90%-efficiency self-balancing ZVS three-
   level DC-DC regulator with constant-frequency AOT V2 control and 5ns ZVS turn-on delay.
   In *Proceedings of the IEEE International Solid-State Circuits Conference (ISSCC)* (pp. 226–
   227). https://doi.org/10.1109/ISSCC.2016.7417989.

8. Barner, A., Wittmann, J., Rosahl, T., & Wicht, B. (2016). A 10 MHz, 48-to-5V synchronous converter with dead time enabled 125 ps resolution zero-voltage switching. In *Proceeding of the IEEE Applied Power Electronics Conference and Exposition (APEC)* (pp. 106–110). https://doi.org/10.1109/APEC.2016.7467859.

9. Wittmann, J., Barner, A., Rosahl, T., & Wicht, B. (2015). A 12V 10MHz buck converter with dead time control based on a 125 ps differential delay chain. In *Proceedings of the ESSCIRC Conference 2015 – 41st European Solid-State Circuits Conference (ESSCIRC)* (pp. 184–187). https://doi.org/10.1109/ESSCIRC.2015.7313859.

10. Wittmann, J., Barner, A., Rosahl, T., & Wicht, B. (2016). An 18 V input 10 MHz buck converter with 125 ps mixed-signal dead time control. *IEEE Journal of Solid-State Circuits, 51*(7), 1705–1715. ISSN:0018-9200. https://doi.org/10.1109/JSSC.2016.2550498.

11. Wittmann, J., Funk, T., Rosahl, T., & Wicht, B. (2018). A 48-V wide-V_{in} 9–25-MHz resonant DC–DC converter. *IEEE Journal of Solid-State Circuits, 53*(7), 1936–1944. ISSN:0018-9200. https://doi.org/10.1109/JSSC.2018.2827953.

12. Funk, T., Wittmann, J., Rosahl, T., & Wicht, B. (2015). A 20 V, 8 MHz resonant DCDC converter with predictive control for 1 ns resolution soft-switching. In *Proceedings of the IEEE International Symposium on Circuits and Systems (ISCAS)* (pp. 1742–1745). https://doi.org/10.1109/ISCAS.2015.7168990.

13. Singh, R. (n.d.). *Silicon carbide switches in emerging applications.* https://www.psma.com/sites/default/files/uploads/tech-forums-semiconductor/presentations/16-silicon-carbide-switches-emerging-applications.pdf.

14. Efficient Power Conversion. (2018). *Where is GaN going?* https://epc-co.com/epc/GalliumNitride/whereisgangoing.aspx.

15. Lobsiger, Y., & Kolar, J. W. (2013). Closed-loop di/dt & dv/dt control and dead time minimization of IGBTs in bridge leg configuration. In *Proceedings of the IEEE 14th Workshop Control and Modeling for Power Electronics (COMPEL)* (pp. 1–7). https://doi.org/10.1109/COMPEL.2013.6626392.

16. Groeger, J., Schindler, A., Wicht, B., & Norling, K. (2017). Optimized dv/dt, di/dt sensing for a digitally controlled slope shaping gate driver. In *Proceedings of the IEEE Applied Power Electronics Conference and Exposition (APEC)* (pp. 3564–3569). https://doi.org/10.1109/APEC.2017.7931209.

17. Oinonen, M., Laitinen, M., & Kyyrä, J. (2014). Current measurement and short-circuit protection of an IGBT based on module parasitics. In *Proceedings of the 16th European Conference Power Electronics and Applications* (pp. 1–9). https://doi.org/10.1109/EPE.2014.6910824.

18. Shalmany, S. H., Draxelmayr, D., & Makinwa, K. A. A. (2013). A micropower battery current sensor with ±0.03% (3σ) inaccuracy from −40 to +85°C. In *Proceedings of the IEEE International Solid-State Circuits Conference Digest of Technical Papers* (pp. 386–387). https://doi.org/10.1109/ISSCC.2013.6487781.

19. Rothan, F., Lhermet, H., Zongo, B., Condemine, C., Sibuet, H., Mas, P., et al. (2011). A ±1.5% nonlinearity 0.1-to-100A shunt current sensor based on a 6kV isolated micro-transformer for electrical vehicles and home automation. In *Proceedings of the IEEE International Solid-State Circuits Conference* (pp. 112–114). https://doi.org/10.1109/ISSCC.2011.5746242.

20. Shalmany, S. H., Draxelmayr, D., & Makinwa, K. A. A. (2017). A ±36-A integrated current-sensing system with a 0.3% gain error and a 400μA offset from −55 °C to +85 °C. In *IEEE Journal of Solid-State Circuits, 52*(4), 1034–1043. ISSN: 0018-9200. https://doi.org/10.1109/JSSC.2016.2639535.

21. Xu, L., Shalmany, S. H., Huijsing, J. H., Makinwa, K. A. A. (2018). A ± 12-A high-side current sensor with 25 V input CM range and 0.35% gain error from −40 °C to 85 °C. *IEEE Solid-State Circuits Letters, 1*(4), 94–97. ISSN:2573-9603. https://doi.org/10.1109/LSSC.2018.2855407.

22. Ziegler, S., Woodward, R. C., Iu, H. H.-C., Borle, L. J. (2009a). Investigation into static and dynamic performance of the copper trace current sense method. *IEEE Sensors Journal, 9*(7), 782–792. ISSN: 1530-437X. https://doi.org/10.1109/JSEN.2009.2021803.

23. Castellanos, J. C., Turhan, M., & Cantatore, E. (2018). A 93.3% peak-efficiency self-resonant hybrid-switched-capacitor LED driver in 0.18μm CMOS technology. *IEEE Journal of*

Solid-State Circuits, 53(7), 1924–1935. ISSN: 0018-9200. https://doi.org/10.1109/JSSC.2018. 2828097.

24. Castellanos, J. C., Turhan, M., Hendrix, M. A. M, van Roermund, A., & Cantatore, E. (2017). A 92.2% peak-efficiency self-resonant hybrid switched-capacitor LED driver in 0.18μm CMOS. In *Proceedings of the ESSCIRC 2017 – 43rd IEEE European Solid State Circuits Conference* (pp. 344–347). https://doi.org/10.1109/ESSCIRC.2017.8094596.

25. Lam, H. Y. H., Ki, W.-H., & Ma, D. (2004). Loop gain analysis and development of high-speed high-accuracy current sensors for switching converters. *Proceedings of the 2004 IEEE International Symposium on Circuits and Systems (IEEE Cat. No.04CH37512), 5*, V. https://doi.org/10.1109/ISCAS.2004.1329936.

26. Jiang, Y., Swilam, M., Asar, S., & Fayed, A. (2018). An accurate sense-FET-based inductor current sensor with wide sensing range for buck converters. In *Proceedings of the IEEE International Symposium on Circuits and Systems (ISCAS)* (pp. 1–4). https://doi.org/10.1109/ ISCAS.2018.8351083.

27. Phan, K. L., Boeve, H., Vanhelmont, F., Ikkink, T., & Talen, W. (2005). Geometry optimization of TMR current sensors for on-chip IC testing. *IEEE Transactions on Magnetics, 41*(10), 3685– 3687. ISSN: 0018-9464. https://doi.org/10.1109/TMAG.2005.854813.

28. Cubells-Beltrán, M. D., Reig, C., Martos, J., Torres, J., & Soret, J. (2011). Limitations of magnetoresistive current sensors in industrial electronics applications. *International Review of Electrical Engineering-IREE, 6*, 423–429.

29. Slatter, R. (2015). High accuracy, high bandwidth magnetoresistive current sensors for spacecraft power electronics. In *Proceedings of the 17th European Conference Power Electronics and Applications (EPE'15 ECCE-Europe)* (pp. 1–10). https://doi.org/10.1109/EPE.2015. 7309419.

30. Slatter, R., & Kramb, M. (2016). High bandwidth, highly integrated current sensors for high power density electromobility applications. In *Proceedings of the 8th IET International Conference on Power Electronics, Machines and Drives (PEMD 2016)* (pp. 1–6). https://doi. org/10.1049/cp.2016.0344.

31. Georgakopoulos, I., Hadjigeorgiou, N., & Sotiriadis, P. P. (2017). A CMOS closed loop AMR sensor architecture. In *Proceedings of the Panhellenic Conference Electronics and Telecommunications (PACET)* (pp. 1–4). https://doi.org/10.1109/PACET.2017.8259975.

32. Snoeij, M. F., Schaffer, V., Udayashankar, S., & Ivanov, M. V. (2016). Integrated fluxgate magnetometer for use in isolated current sensing. *IEEE Journal of Solid-State Circuits, 51*(7), 1684–1694. ISSN: 0018-9200. https://doi.org/10.1109/JSSC.2016.2554147.

33. Snoeij, M. F., Schaffer, V., Udayashankar, S., & Ivanov, M. V. (2015). An integrated fluxgate magnetometer for use in closed-loop/open-loop isolated current sensing. In *Proceedings of the ESSCIRC Conference 2015 – 41st European Solid-State Circuits Conference (ESSCIRC)* (pp. 263–266). https://doi.org/10.1109/ESSCIRC.2015.7313877.

34. Kashmiri, M., Kindt, W., Witte, F., Kearey, R., & Carbonell, D. (2015). A 200kS/s 13.5b integrated-fluxgate differential-magnetic-to-digital converter with an oversampling compensation loop for contactless current sensing. In *Proceedings of the IEEE International Solid-State Circuits Conference – (ISSCC) Digest of Technical Papers* (pp. 1–3). https://doi.org/10.1109/ ISSCC.2015.7063140.

35. Ramsden, E. (2011). *Hall-effect sensors: theory and application.* Amsterdam: Elsevier.

36. Heidari, H., Bonizzoni, E., Gatti, U., & Maloberti, F. (2015a). A CMOS current-mode magnetic hall sensor with integrated front-end. *IEEE Transactions on Circuits and Systems I: Regular Papers, 62*(5), 1270–1278. ISSN: 1549-8328. https://doi.org/10.1109/TCSI.2015.2415173.

37. Jiang, J., Kindt, W. J., & Makinwa, K. A. A. (2014). A continuous-time ripple reduction technique for spinning-current hall sensors. *IEEE Journal of Solid-State Circuits, 49*(7), 1525– 1534. ISSN: 0018-9200. https://doi.org/10.1109/JSSC.2014.2319252.

38. Jiang, J., & Makinwa, K. A. A. (2015). A multi-path CMOS hall sensor with integrated ripple reduction loops. In *Proceedings of the IEEE Asian Solid-State Circuits Conference (A-SSCC)* (pp. 1–4). https://doi.org/10.1109/ASSCC.2015.7387504.

39. Jiang, J., & Makinwa, K. (2016a). A hybrid multi-path CMOS magnetic sensor with 76 ppm/ °C sensitivity drift. In *Proceedings of the ESSCIRC Conference 2016: 42nd European Solid-State Circuits Conference* (pp. 397–400). https://doi.org/10.1109/ESSCIRC.2016.7598325.

40. Jiang, J., & Makinwa, K. A. A. (2017a). A hybrid multi-path CMOS magnetic sensor with 76 ppm/ °C sensitivity drift and discrete-time ripple reduction loops. *IEEE Journal of Solid-State Circuits, 52*(7), 1876–1884. ISSN: 0018-9200. https://doi.org/10.1109/JSSC.2017.2685462.

41. Blanchard, H., de Raad Iseli, C., & Popovic, R. S. (1997). Compensation of the temperature-dependent offset drift of a hall sensor. *Sensors and Actuators A: Physical, 60*(1), 10–13. ISSN: 0924-4247. http://www.sciencedirect.com/science/article/pii/S0924424796014112.

42. Innosen (Rev2.0. 4.140108). Temperature compensation bipolar hall effect sensor. In *Datasheet ES413/ES513*.

43. Micronas (Edition Dec. 8, 2008). Linear hall-effect sensor IC. In *Datasheet HAL 411*.

44. Instruments, Texas (2018). Automotive ratiometric linear Hall effect sensor. In *Datasheet DRV5055-Q1*.

45. Karrer, N., & Hofer-Noser, P. (1999). A new current measuring principle for power electronic applications. In *Proceedings of the (Cat. No.99CH36312) 11th International Symposium on Power Semiconductor Devices and ICs. ISPSD'99* (pp. 279–282). https://doi.org/10.1109/ISPSD.1999.764117.

46. Dalessandro, L., Karrer, N., & Kolar, J. W. (2007). High-performance planar isolated current sensor for power electronics applications. In *IEEE Transactions on Power Electronics, 22*(5), 1682–1692. ISSN: 0885-8993. https://doi.org/10.1109/TPEL.2007.904198.

47. Karrer, N., Hofer-Noser, P., & Henrard, D. (1999). HOKA: a new isolated current measuring principle and its features. In *Proceedings of the Conference Record of the 1999 IEEE Industry Applications Conference. Thirty-Forth IAS Annual Meeting (Cat. No.99CH36370)* (Vol. 3, pp. 2121–2128). https://doi.org/10.1109/IAS.1999.806028.

48. Jiang, J., & Makinwa, K. (2016b). A hybrid multipath CMOS magnetic sensor with 210μTrms resolution and 3MHz bandwidth for contactless current censing. In *Proceedings of the IEEE International Solid-State Circuits Conference (ISSCC)* (pp. 204–205). https://doi.org/10.1109/ISSCC.2016.7417978.

49. Jiang, J., & Makinwa, K. A. A. (2017b). Multipath wide-bandwidth CMOS magnetic sensors. In *IEEE Journal of Solid-State Circuits, 52*(1), 198–209. ISSN: 0018-9200. https://doi.org/10.1109/JSSC.2016.2619711.

50. Paun, M.-A., Sallese, J.-M., & Kayal, M. (2013a). Hall effect sensors design, integration and behavior analysis. *Journal of Sensor and Actuator Networks, 2*, 85–97.

51. Paun, M.-A., Sallese, J.-M., & Kayal, M. (2013b). Comparative study on the performance of five different hall effect devices. In *Sensors 13.2* (pp. 2093–2112).

52. Kaufmann, T., Purkl, F., Ruther, P., & Paul, O. (2011). Novel coupling concept for five-contact vertical hall devices. In *Proceedings of the 16th International Solid-State Sensors, Actuators and Microsystems Conference* (pp. 2855–2858). https://doi.org/10.1109/TRANSDUCERS.2011.5969126.

53. Sung, G., Wang, W., & Yu, C. (2017). Analysis and modeling of one-dimensional folded vertical hall sensor with readout circuit. *IEEE Sensors Journal, 17*(21), 6880–6887. ISSN: 1530-437X. https://doi.org/10.1109/JSEN.2017.2754295.

54. Heidari, H., Bonizzoni, E., Gatti, U., Maloberti, F., & Dahiya, R. (2015b). Optimal geometry of CMOS voltage-mode and current-mode vertical magnetic hall sensors. In *Proceedings of the IEEE SENSORS* (pp. 1–4). https://doi.org/10.1109/ICSENS.2015.7370365.

55. Sung, G.-M., Gunnam, L. C., Wang, H.-K., & Lin, W.-S. (2018). Three-dimensional CMOS differential folded hall sensor with bandgap reference and readout circuit. In *IEEE Sensors Journal, 18*(2), 517–527. ISSN: 1530-437X. https://doi.org/10.1109/JSEN.2017.2777485.

56. Paranjape, M., Ristic, L., & Filanovsky, I. (1991). A 3-D vertical hall magnetic field sensor in CMOS technology. In *Proceedings of the TRANSDUCERS '91: 1991 International Conference Solid-State Sensors and Actuators. Digest of Technical Papers* (pp. 1081–1084). https://doi.org/10.1109/SENSOR.1991.149085.

57. Sander, C., Leube, C., & Paul, O. (2015). Novel compact two-dimensional CMOS vertical hall sensor. In *Proceedings of the Actuators and Microsystems (TRANSDUCERS) 2015 Transducers – 2015 18th International Conference Solid-State Sensors* (pp. 1164–1167). https://doi.org/10.1109/TRANSDUCERS.2015.7181135.
58. Rogowski, W., & Steinhaus, W. (1912). Die Messung der magnetischen Spannung. *Archiv für Elektrotechnik, 1*(4), 141. https://doi.org/10.1007/BF01656479.
59. Ziegler, S., Woodward, R. C., Iu, H. H.-C. Borle, L. J. (2009b). Current sensing techniques: a review. *IEEE Sensors Journal, 9*(4), 354–376. ISSN: 1530-437X. https://doi.org/10.1109/JSEN.2009.2013914.
60. Pascal, J. et al. (Mar. 2012). "Electronic Front End for Rogowski Coil Current Transducers with Online Accuracy Self Monitoring". In: *Proc. IEEE Int. Conf. Industrial Technology*, pp. 1037–1040. https://doi.org/10.1109/ICIT.2012.6210076.
61. Xue, Y., Bloch, R., Isler, S., & Georges, L. (2014). A compact planar Rogowski coil current sensor for active current balancing of parallel-connected silicon carbide MOSFETs. In *Proceedings of the IEEE Energy Conversion Congress and Exposition (ECCE)* (pp. 4685–4690). https://doi.org/10.1109/ECCE.2014.6954042.
62. Han, R., Wu, J.-W., Ding, W.-D., Jing, Y., Zhou, H.-B., Liu, Q.-J. et al., (2015). Hybrid PCB Rogowski coil for measurement of nanosecond-risetime pulsed current. In *IEEE Transactions on Plasma Science, 43*(10), 3555–3561. ISSN: 0093-3813. https://doi.org/10.1109/TPS.2015.2415517.
63. Wang, B., Wang, D., & Wu, W. (2009). A Rogowski coil current transducer designed for wide bandwidth current pulse measurement. In *Proceedings of the IEEE 6th International Power Electronics and Motion Control Conference* (pp. 1246–1249). https://doi.org/10.1109/IPEMC.2009.5157575.
64. Limcharoen, W., & Yutthagowith, P. (2013). Rogowski coil with an active integrator for measurement of switching impulse current. In *Proceedings of the Telecommunications and Information Technology 2013 10th International Conference Electrical Engineering/Electronics, Computer* (pp. 1–4). https://doi.org/10.1109/ECTICon.2013.6559578.
65. Yutthagowith, P., & Leelachariyakul, B. (2014). A Rogowski coil with an active integrator for measurement of long duration impulse currents. In *Proceedings of the International Conference on Lightning Protection (ICLP)* (pp. 1761–1765). https://doi.org/10.1109/ICLP.2014.6973414.
66. Liu, Y., Lin, F., Zhang, Q., & Zhong, H. (2011). Design and construction of a Rogowski coil for measuring wide pulsed current. *IEEE Sensors Journal, 11*(1), 123–130. ISSN: 1530-437X. https://doi.org/10.1109/JSEN.2010.2052034.
67. Ray, W. F., & Hewson, C. R. (2000). High performance Rogowski current transducers. In *Proceedings of the Conference Record of the 2000 IEEE Industry Applications Conference. Thirty-Fifth IAS Annual Meeting and World Conference on Industrial Applications of Electrical Energy (Cat. No.00CH37129)* (Vol. 5, pp. 3083–3090). https://doi.org/10.1109/IAS.2000.882606.
68. Zhao, L., van Wyk, J. D., & Odendaal, W. G. (2004). Planar embedded pick-up coil sensor for power electronic modules. In *Proceedings of the APEC '04. Nineteenth Annual IEEE Applied Power Electronics Conference and Exposition* (Vol. 2, pp. 945–951). https://doi.org/10.1109/APEC.2004.1295936.
69. Ho, G. K. Y., Fang, Y., Pong, B. M. H., & Hui, R. S. Y. (2017). Printed circuit board planar current transformer for GaN active diode'. In *Proceedings of the IEEE Applied Power Electronics Conference and Exposition (APEC)* (pp. 2549–2553). https://doi.org/10.1109/APEC.2017.7931056.
70. Guillod, T., Gerber, D., Biela, J., & Muesing, A. (2012). Design of a PCB Rogowski coil based on the PEEC method. In *Proceedings of the 7th International Conference on Integrated Power Electronics Systems (CIPS)* (pp. 1–6).
71. Wang, K., Yang, X., Li, H., Wang, L., & Jain, P. (2018). A high-bandwidth integrated current measurement for detecting switching current of fast GaN devices. *IEEE Transactions on Power Electronics, 33*(7), 6199–6210. ISSN: 0885-8993. https://doi.org/10.1109/TPEL.2017.2749249.

72. Qing, C., Hong-bin, L., Ming-ming, Z., & Yan-bin, L. (2006). Design and characteristics of two Rogowski coils based on printed circuit board. *IEEE Transactions on Instrumentation and Measurement, 55*(3), 939–943. ISSN: 0018-9456. https://doi.org/10.1109/TIM.2006.873788

73. Tröster, N., Wölfle, J., Ruthardt, J., & Roth-Stielow, J. (2017). High bandwidth current sensor with a low insertion inductance based on the HOKA principle. In *Proceedings of the 19th European Conference Power Electronics and Applications (EPE'17 ECCE Europe)* (pp. P.1– P.9). https://doi.org/10.23919/EPE17ECCEEurope.2017.8099003.

74. Nibir, S. J., Hauer, S., Biglarbegian, M., & Parkhideh, B. (2018). Wideband contactless current sensing using hybrid Magnetoresistor-Rogowski sensor in high frequency power electronic converters. In *Proceedings of the IEEE Applied Power Electronics Conference and Exposition (APEC)* (pp. 904–908). https://doi.org/10.1109/APEC.2018.8341121.

Chapter 3
Rogowksi Coil Current Sensor

The Rogowski coil is well suitable for sensing fast changing signals currents. It provides high bandwidth and galvanic isolation for power electronics applications, as derived in Sect. 2.2.6.

In this book, Rogowski coil based current sensors for the on-chip and off-chip current sensing are proposed. Section 3.1 shows the operation principle of a Rogowski coil and the resulting dependencies for the design of the coil in terms of bandwidth and sensitivity. The design of a planar coil for on-chip current sensing is shown in Sect. 3.2 [1], while Sect. 3.3 covers the design of a helix-shaped coil for off-chip current sensing [2].

3.1   Operation Principle

A Rogowski coil measures the change of magnetic field $\vec{B}$ of the current to be measured I_{meas}, based on Faraday's law. This implies that the induced voltage in a loop is proportional to the rate of change over time of $\vec{B}$, respectively, I_{meas}. The induced voltage V_o of a Rogowski coil can be calculated by Xiang et al. [3]

$$V_o(t) = \frac{d\phi}{dt} = M \cdot \frac{dI_{meas}}{dt}, \tag{3.1}$$

while I_{meas} is the current to be measured though a power line and M is the mutual inductance between the power line and coil. M depends on geometry, dimensions, and number of windings of the coil, as indicated in Fig. 3.1. According to Eq. 3.1, the induced voltage V_o has no bandwidth limitation. High signal frequencies even enhance V_o, because it is proportional to the signal frequency.

Figure 3.2 shows the equivalent circuit of a Rogowski coil based on a Lumped parameter model (LPM). This model is valid up to the first resonance frequency

© The Editor(s) (if applicable) and The Author(s), under exclusive license to
Springer Nature Switzerland AG 2020
T. Funk, B. Wicht, *Integrated Wide-Bandwidth Current Sensing*,
https://doi.org/10.1007/978-3-030-53250-5_3

Fig. 3.1 Principle of the
Rogowski coil for current
sensing

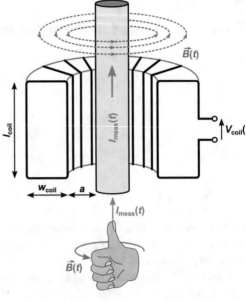

Fig. 3.2 Equivalent circuit of
a Rogowski coil

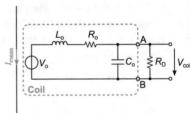

of the coil [3, 4]. For higher frequencies, the coil must be divided into several
parts, which can be described by a Distributed parameter model (DPM)[3] or Partial
element equivalent circuit (PEEC)[4]. The resonant frequency of the integrated
Rogowski coil in this work is $\gg 100\,\text{MHz}$. Thus, consideration of the transfer
behavior up to the first pole is sufficient and the LPM is valid.

The equivalent circuit of the LPM consists of a voltage source V_o, a parasitic
inductance L_o, a parasitic resistance R_o, and a parasitic capacitance C_o. L_o and C_o
limit the bandwidth of the Rogowski coil and R_o forms a voltage divider with the
required damping resistance R_D. Due to these parasitic components, the Rogowski
coil has a second order transfer function $R(s)$, which is given as

$$R(s) = \frac{V_{\text{coil}}(s)}{I_{\text{meas}}(s)} = \frac{M R_D s}{R_D C_o L_o s^2 + (R_D R_o C_o + L_o)s + R_o + R_D}. \tag{3.2}$$

This transfer behavior can also be represented in the standard form

$$R(s) = \frac{M R_D s}{\frac{1}{\omega_o^2}s^2 + \frac{2D}{\omega_o}s + 1}. \tag{3.3}$$

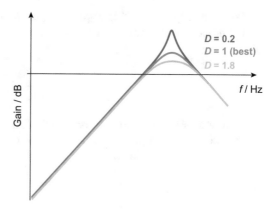

Fig. 3.3 Transfer behavior of a Rogowski coil for different damping coefficients

By coefficient comparison between Eqs. 3.2 and 3.3 the natural frequency ω_o can be determined as

$$\omega_o = \sqrt{\frac{R_o + R_D}{C_o L_o R_D}},$$ (3.4)

and the damping coefficient D as

$$D = \frac{\omega_o}{2} \cdot \frac{R_D R_o C_o + L_o}{R_D + R_o}.$$ (3.5)

Depending on the value of D the transfer behavior close to the natural frequency ω_o is changing. For $0 < D < 1$, there are overshoots possible due to the parasitic inductance L_o and parasitic capacitance C_o. If $D > 1$, there is no overshoot, but the dominant pole of the coil goes to lower frequencies and the bandwidth is reduced [5]. The best transfer behavior can be achieved by $D = 1$, as shown in Fig. 3.3.

The damping coefficient D depends on the parasitics of the Rogowski coil and the damping resistance R_D. Therefore, the value of R_D can be used to set $D = 1$ for different coils. Inserting Eq. 3.4 into 3.5 and solving for R_D results in

$$R_D = -L_o \frac{2D\sqrt{C_o \left(R_o{}^2 D^2 C_o - R_o{}^2 C_o + L_o\right)} + R_o C_o - 2R_o D^2 C_o}{R_o{}^2 C_o{}^2 - 4D^2 L_o C_o}.$$ (3.6)

For $D = 1$ the equivalent circuit has a double pole at f_o, which defines the maximum bandwidth BW_{Rog} of the Rogowski coil.

$$f_o = BW_{Rog} = \frac{\omega_o}{2\pi}$$ (3.7)

For frequency components $f_{meas} \ll f_o$, the output voltage V_{coil} of the Rogowski coil is proportional to the mutual inductance M and the voltage divider of R_o and R_D.

$$V_{coil} \propto M \cdot \frac{R_D}{R_o + R_D} \tag{3.8}$$

Therefore, M is a measure of sensitivity S_{Rog} and depends on the design of the Rogowski coil. The larger the M, the larger the output signal V_{coil} at a specific signal frequency f_{meas} of the current I_{meas} to be measured.

In conclusion, the transfer function Eq. 3.2 and the transfer behavior in Fig. 3.3 indicate that the Rogowski coil has a differentiating transfer behavior. Therefore, the output voltage V_{coil} is proportional to the derivative dI/dt of the current to be measured I_{meas} and increases with $+20\,dB/decade$. The mutual inductance M is a measure for the magnetic coupling between the power line and the Rogowski coil and depends on design of the coil. The bandwidth BW_{Rog} of the Rogowski coil results from the parasitic components of the coil itself. The best transfer behavior and highest bandwidth BW_{Rog} can be achieved for the damping coefficient $D = 1$. D depends on the parasitic components (R_o, L_o, and C_o) of the coil and is adjusted by the damping resistor R_D.

3.2 Rogowski Coil for On-Chip Current Sensing

The need for a fully integrated current sensor with a bandwidth in the range 50 MHz to 100 MHz was derived in Sect. 2.1.1. To use the properties of a Rogowski coil for on-chip current sensing most effectively, a planar Rogowski coil is placed besides an on-chip power line and measures the magnetic field $\vec{B}$ in the z-direction, as shown in Fig. 3.4. This allows an area-efficient current sensor with high bandwidth BW_{Rog} and high sensitivity S_{Rog}[1].

The proposed concept is established for discrete Printed circuit board (PCB) implementations, but with a limited bandwidth <30 MHz and large area [6–8]. Due to the integrated implementation, the distance between the coil and power line can be minimized to a few μm. This leads to higher sensitivity S_{Rog}, since the magnetic

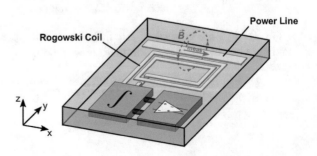

Fig. 3.4 On-chip current sensing with an integrated Rogowski coil

field $\vec{B}$ is inversely proportional to the distance to the power line. Furthermore, by considering the parasitics of an integrated planar Rogowski coil, a bandwidth of >1.5 GHz is achieved by an optimized design for a small area of 0.35 mm². With an optimized layout of the coil, the effects of parasitic capacitive coupling can be improved by a factor of 100 and current sensing with superior characteristics is possible.

There are many degrees of freedom for the design of the planar coil. The area determines the cross-section of the coil and maximum number and size of windings. Therefore, for the design of the planar coil, the available area of the coil is the first value to be defined. In principle, the coil should be long rather than wide ($l_{coil} > w_{coil}$ in Fig. 3.1) to achieve a high sensitivity. Thus, in the following considerations, the area is set to a maximum size of 700 µm × 500 µm as an example. All comparisons for the design of the planar coil in following sections refer to this area.

Section 3.2.1 shows how the magnetic coupling and thus the sensitivity S_{Rog} of a planar integrated Rogowski coil can be calculated and optimized. In Sect. 3.2.2 the relevant parasitic components of the coil are calculated and their dependencies for different designs are shown. The resulting design trade-offs for bandwidth and sensitivity are discussed in Sect. 3.2.3 and verified with 3D field simulations in Sect. 3.2.4. The influence of voltage coupling due to parasitic capacitances in the sensor element, and how this can be improved, is described in Sect. 3.2.5.

3.2.1 Mutual Inductance

The mutual inductance M represents the magnetic coupling between the power line and the Rogowski coil [3–5]. M depends on dimensions and number of windings of the coil and the distance to the power line. With a square layout of the coil, the largest area of windings can be realized for the given area. Figure 3.5 shows the

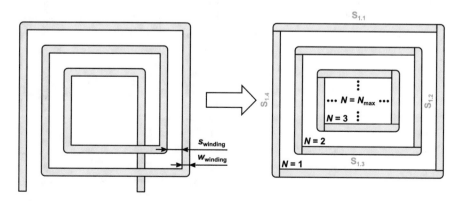

Fig. 3.5 Geometry model of planar Rogowski coil layout for mutual inductance M_p calculation

geometry model of a planar coil with multiple windings for the calculation of M [9, 10]. With this approach, the optimum number of windings N, winding width $w_{winding}$, and space $s_{winding}$ between the windings of a planar Rogowski coil can be calculated.

The total mutual inductance M_p of the planar coil is the summation of the mutual inductances of all windings and is defined as

$$M_p = \sum_{N=1}^{N=N_{max}} M_{p,N} \tag{3.9}$$

for a coil with N_{max} windings [11]. While, the mutual inductance $M_{p,N}$ of the winding N is defined as

$$M_{p,N} = \frac{\mu_o \cdot l_{coil,N}}{2\pi} \cdot \ln\left(\frac{g_{outer}}{g_{inner}}\right). \tag{3.10}$$

$M_{p,N}$ is proportional to the length of the segments $l_{coil,N}$ and depends on the Geometric mean distance (GMD) g_{inner} and g_{outer} between the power line and the coil. For an integrated coil, the GMD must be considered, since the width of the power line $w_{powerline}$ and the width of a winding $w_{winding}$ cannot be neglected compared to discrete Rogowski coils. In discrete implementations, the width of these wires can be assumed as negligibly thin [11–14]. The simplified coil of winding $N = 1$ and the power line is shown in Fig. 3.6.

Figure 3.7 shows two wires with different widths $w_{powerline}$ and $w_{winding}$ for the calculation of GMD. The power line has the leftmost point in the origin and the coil has an offset by a distance c. Point P_y is located at a distance y from the origin in the power line, and P_x is a point located at a distance x from the origin in the coil segment $S_{1.1}$, respectively, $S_{1.3}$ of the coil. The calculation of the GMD for wires

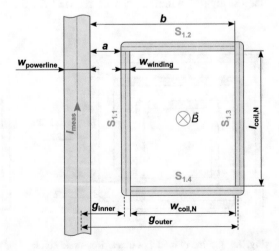

Fig. 3.6 Geometry of single turn Rogowski coil for mutual inductance M_p calculation

Fig. 3.7 Dimensions for the calculation of the GMD between two lines with different lengths

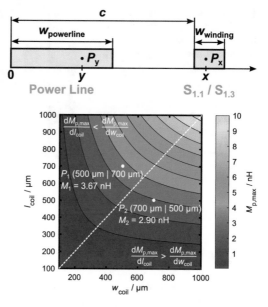

Fig. 3.8 Maximum mutual inductances $M_{p,max}$ for different size of the planar Rogowski coil

with the same width was published in [15, 16]. Equation 3.11 is based on these equations, but adjusted for wires with different widths, since $w_{powerline} \gg w_{winding}$.

The logarithm of GMD is calculated by the sum of the logarithms of the distance between every combination of points $(x - y)$ and divided by $w_{powerline}$ and $w_{winding}$:

$$\log(g) = \frac{1}{w_{powerline} \cdot w_{winding}} \int_c^{c+w_{winding}}$$

$$\times \int_0^{w_{powerline}} \log(x - y) \mathrm{d}x \mathrm{d}y. \tag{3.11}$$

The distance c is different for g_{inner} and g_{outer}, and is defined by Eq. 3.12.

$$c = \begin{cases} w_{powerline} + a & \text{for } g_{inner} \\ w_{powerline} + b & \text{for } g_{outer} \end{cases} \tag{3.12}$$

To calculate M_p by Eqs. 3.9 and 3.10, the GMD g_{inner} and g_{outer} and the length $l_{coil,N}$ of segments $S_{N.1}$ and $S_{N.3}$ must be calculated for each winding by Eq. 3.11.

The maximum mutual inductance $M_{p,max}$ for different size of the coil is shown in Fig. 3.8. As expected from Eq. 3.10 to 3.12, $M_{p,max}$ strongly depends on the available area and increases with an increasing l_{coil} and w_{coil}. Depending on the ratio of l_{coil}/w_{coil}, it is more efficient to optimize l_{coil} or w_{coil}, as suggested in Fig. 3.8. For example, if l_{coil} is greater than w_{coil} (at left side of dotted line in Fig. 3.8), the change of $M_{p,max}$ by changing w_{coil} is greater than the change of l_{coil}.

Fig. 3.9 Mutual inductances M_p of a planar Rogowski coil for different number of windings N and winding widths $w_{winding}$

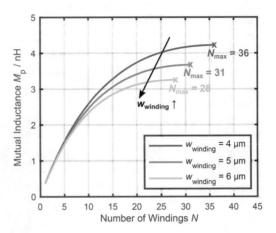

Generally, for a constant area, a higher mutual inductance M_p can be reached by $l_{coil} > w_{coil}$, e.g., for an area of $0.35\,\text{mm}^2$, $M_{p,max}$ of point P_1 is higher than that of P_2 ($3.67\,\text{nH} > 2.90\,\text{nH}$).

Figure 3.9 shows the total mutual inductance M_p for different numbers of windings for a given coil area of $l_{coil} \times w_{coil} = 700 \times 500\,\mu\text{m}$ (P_1). As expected from Eq. 3.9, M_p increases with an increasing number of winding N. For a higher number of windings, there is a saturation of M_p because the area of the inner windings becomes smaller and the contribution of these windings is lower than the larger outer windings.

The mutual inductance M_p and the maximum number of windings N_{max} decrease with the increasing winding width $w_{winding}$, since the available area of each winding is getting smaller and number of maximal possible windings decreases. The highest value of M_p is achieved at the smallest wire width $w_{winding} = 4\,\mu\text{m}$ and $N = 36$ windings.

As a conclusion, the mutual inductance M_p depends on the area of each winding and the total number of possible windings of the coil. To achieve a large M_p, the wire width $w_{winding}$ and consequently the space $s_{winding}$ must be minimal to get the highest number of possible windings N_{max} with the largest area of each winding. Generally, the larger the area of the coil, the higher the M_p.

Up to now, the influence of parasitic components, such as resistance R_o, inductance L_o, and capacitances C_o, have been neglected. These components must be considered for the transfer behavior of the Rogowski coil, since they will limit the bandwidth BW_{Rog} and reduce the sensitivity S_{Rog} of the coil.

3.2.2 Parasitic Components

The parasitic components of the coil, such as resistance R_o, inductance L_o, and capacitances C_o, have a strong impact on the transfer behavior of the Rogowski coil. While L_o and C_o will limit the bandwidth BW_{Rog}, R_o will degrade the sensitivity S_{Rog} of the sensor. These parasitic components must be as small as possible to achieve the highest performance of the Rogowski coil. They are depending on the dimensions and the number of windings of the coil. Therefore, only with an optimized design of the coil, a high bandwidth BW_{Rog} with a high sensitivity S_{Rog} can be achieved at the same time.

In the following these parasitic components are investigated for different coil designs and for the given area ($700 \times 500\,\mu m$). This allows to derive subsequently the optimal design of the Rogowski coil for a high bandwidth with a simultaneous high sensitivity.

Parasitic Resistance R_o

The parasitic resistance R_o forms a voltage divider with the required damping resistance R_D, see Fig. 3.2. In order to achieve a low voltage drop over R_o and thus a high sensitivity of the Rogowski coil, it must be as small as possible. R_o is proportional to the total length $l_{coil,tot}$ of all segments of the coil and the sheet resistance R_S of the used metal layer and depends on the winding width $w_{winding}$:

$$R_o = R_S \cdot \frac{l_{coil,tot}}{w_{winding}} \tag{3.13}$$

R_o for an increasing number of windings N and different winding widths $w_{winding}$ is shown in Fig. 3.10. The slopes increase not linearly, since the size of the inner windings gets smaller and, as expected from Eq. 3.13, R_o decreases with the winding width $w_{winding}$.

Finally, for a high sensitivity S_{Rog} of the coil, R_o must be as low as possible. To achieve a low R_o, the winding width $w_{winding}$ must be large and the number of windings N low. However, M_p will decrease at the same time. Therefore, there is a design trade-off between the resistance R_o and the mutual inductance M_p.

Parasitic Inductance L_o

The inductance L_o (Fig. 3.2) limits the bandwidth BW_{Rog} of the Rogowski coil. To achieve a high bandwidth, L_o must be as low as possible. It can be divided into three parts, the self-inductances L_S, the sum of all positive mutual inductances M_+, and the sum of all negative mutual inductances M_- between the windings and is defined as

Fig. 3.10 Parasitic resistance R_o of a planar Rogowski coil for different number of windings N and winding widths $w_{winding}$

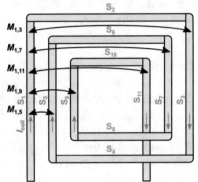

Fig. 3.11 Segments of a planar Rogowski coil for inductances L_o calculation

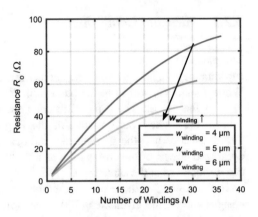

$$L_o = L_S + M_+ - M_- \tag{3.14}$$

for planar coils [17–19]. The self-inductance L_S of a rectangular wire can be calculated by Grover's formula

$$L_S = \frac{\mu_o \cdot l_{winding}}{2\pi} \left[\ln\left(\frac{2l_{winding}}{w_{winding} + t_{winding}}\right) + 0.50049 \right.$$

$$\left. + \frac{w_{winding} + t_{winding}}{3l_{winding}} \right], \tag{3.15}$$

where $l_{winding}$ is the total length of all windings of the coil and $t_{winding}$ is the thickness of the windings [20]. L_S increases with an increasing length $l_{winding}$ and with a decreasing width $w_{winding}$ and / or thickness $t_{winding}$.

For the calculation of the mutual inductances M_+, respectively, M_- between the windings and the inductance L_o, the planar coil must be subdivided into individual segments similar to Sect. 3.2.1. The geometry itself will not be simplified, as shown in Fig. 3.11. If the current I_{coil} of two parallel segments flows in the same direction,

Fig. 3.12 Dimensions for the calculation of mutual inductance M_+ or M_- between two segments of the planar Rogowski coil

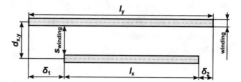

e.g., in segments S_1 and S_5, the mutual inductance $M_{1,5}$ is positive. If the current flows in opposite direction, e.g., in segment S_1 and S_3, the mutual inductance $M_{1,3}$ is negative. Since the parallel segments x and y have the same width w_{winding}, the mutual inductance $M_{x,y}$ can be calculated according to Greenhouse [17]

$$M_{x,y} = 0.5\big[M(l_x + \delta_1) + M(l_x + \delta_2) - \big(M(\delta_1) + M(\delta_2)\big)\big], \tag{3.16}$$

where

$$M(l) = \frac{\mu_0}{2\pi}l\left\{\ln\left[\frac{l}{g} + \sqrt{1 + \left(\frac{l}{g}\right)^2}\right] - \sqrt{1 + \left(\frac{g}{l}\right)^2} + \frac{g}{l}\right\}. \tag{3.17}$$

For segments with the same width w_{winding}, GMD g can be calculated as

$$g = \exp\left(\ln\left(d_{x,y}\right) - \frac{w_{\text{winding}}^2}{12d_{x,y}^2} - \frac{w_{\text{winding}}^4}{60d_{x,y}^4}\right.$$
$$\left. - \frac{w_{\text{winding}}^6}{168d_{x,y}^6} - \frac{w_{\text{winding}}^8}{360d_{x,y}^8} - \frac{w_{\text{winding}}^{10}}{660d_{x,y}^{10}}\right), \tag{3.18}$$

where $d_{x,y}$ is the distance of center between segment x and y. In the calculation of $M_{x,y}$ the different lengths of the respective segments, as shown in Figs. 3.11 and 3.12, must be considered.

Figure 3.13 shows the partition of the inductance L_0 of Eq. 3.14 for different numbers of windings N. The most dominant part of L_0 is the positive mutual inductance M_+ between the windings. This proportion is lowered by M_-, but is still greater than self-inductance L_S. M_+ can be reduced by an increasing space s_{winding} between the windings or an increasing width w_{winding} of the windings. Both result in an increasing distance $d_{x,y}$ between the respective segments, see Fig. 3.12. Additionally, for larger s_{winding} and / or w_{winding} the negative mutual inductance M_- also increases, whereby L_0 further decreases.

The inductance L_0 for different winding widths w_{winding} and an increasing number of windings N is shown in Fig. 3.14. With an increasing w_{winding} the inductance L_0 can be significantly reduced for a large number of windings. This is due to the increasing distance $d_{x,y}$ between the respective segments. For a low number of windings, the difference is smaller, as expected from the break-down of

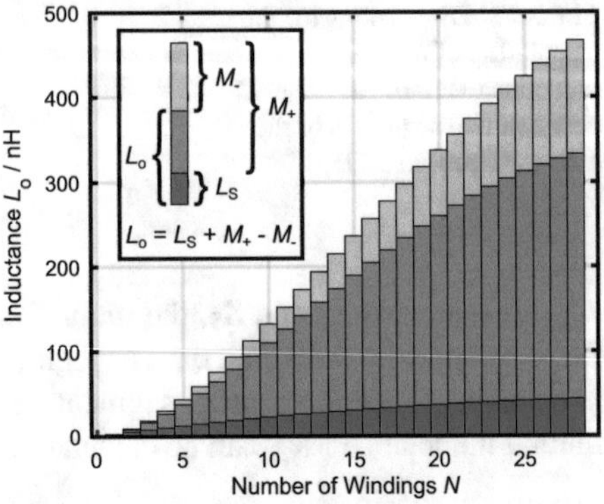

Fig. 3.13 Proportion of the inductance L_o of a planar Rogowski coil for different number of windings N

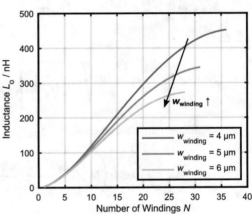

Fig. 3.14 Inductance L_o of a planar Rogowski coil for different number of windings N and winding widths $w_{winding}$

L_o in Fig. 3.13. Here, the mutual inductances M_+ and M_- are lower and the self-inductance L_S has a larger impact.

Generally, the lower the number of windings N, the lower the inductance L_o, and therefore the higher the resulting bandwidth BW_{Rog} of the Rogowski coil. On the one hand, the number of windings and the dimensions of the coil must be large to achieve a high sensitivity S_{Rog} (high M_p), but on the other hand these parameters must be low to achieve a high bandwidth BW_{Rog}. Therefore, there is a trade-off between the mutual inductance M_p and the inductance L_o, as will be investigated in Sect. 3.2.3.

Parasitic Capacitance C_o

The parasitic capacitance C_o also limits the bandwidth BW_{Rog} of the Rogowski coil. To achieve a high bandwidth, C_o must be as low as possible. C_o depends on the substrate capacitance C_{sub} and the overlap capacitance $C_{overlap}$ between the top metal layer (used for the planar Rogowski coil) and the underlying layer for the connection of the coil, as shown in Fig. 3.15.

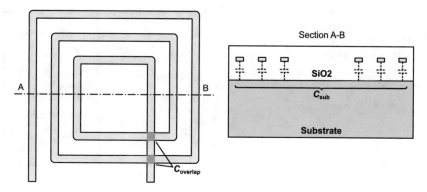

Fig. 3.15 Parasitic capacitances of a planar Rogowski coil

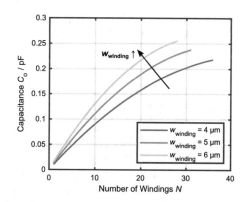

Fig. 3.16 Parasitic capacitance C_o of a planar Rogowski coil for different number of windings N and winding widths $w_{winding}$

In the equivalent circuit of an integrated coil, the total substrate capacitance

$$C_{sub} = \epsilon_{ox} \cdot \frac{l_{coil,tot} \cdot w_{winding}}{t_{ox}} \tag{3.19}$$

is split into half and connected between terminals of the coil and the substrate, see Appendix [21, 22]. For the calculation of C_o these two capacitances are connected in series, therefore C_{sub} is multiplied by the factor of 0.25 and is defined as

$$C_o = 0.25 \cdot C_{sub} + C_{overlap}. \tag{3.20}$$

Figure 3.16 shows C_o for different winding widths $w_{winding}$ and an increasing number of windings N. As expected from Eq. 3.19, C_o increases with the number of windings N and the width $w_{winding}$.

Similar to the inductance L_o, with an increasing C_o, the bandwidth of the Rogowski coil will be limited. To minimize the including substrate capacitance C_{sub} the winding width $w_{winding}$ must be small and the number of windings N low. A small $w_{winding}$ has a positive effect on the mutual inductance M_p, but M_p

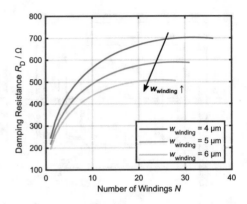

Fig. 3.17 Damping resistor R_D of a planar Rogowski coil for different number of windings N and winding widths $w_{winding}$

decreases due to the low number of windings N. Therefore, there is an additional design trade-off between the capacitance C_o and the mutual inductance M_p, as will be investigated in Sect. 3.2.3.

3.2.3   Design Trade-Offs

The major design goal of a Rogowski coil is a high bandwidth with high sensitivity at the same time. The previous sections revealed a high degree of freedom in the design of a Rogowski coil. There are several trade-offs between the mutual inductance M_p and the parasitic components of the Rogowski coil, as suggested in Sect. 3.2.2. Finally, all these considerations have an impact on the bandwidth BW_{Rog} and the sensitivity S_{Rog} of the Rogowski coil. These correlations are now compared to each other to get an optimal design of the Rogowski coil for on-chip current sensing.

To achieve the best transfer behavior with $D = 1$ (see Sect. 3.1), the damping resistor R_D must be adjusted for every geometric change of the Rogowski coil and can be calculated by Eq. 3.6. Figure 3.17 shows R_D for different winding widths $w_{winding}$ and for an increasing number of windings N. For the area of the proposed coil ($700 \times 500\,\mu m$), R_D is in the range between $194\,\Omega$ and $702\,\Omega$. For larger geometries of the coil, e.g., $1000 \times 1000\,\mu m$, R_D can increase up to $900\,\Omega$. Nevertheless, the resistance of R_D is in a range that allows an integrated implementation.

In the following, the voltage divider between the parasitic resistance R_o and the damping resistance R_D will be considered. One would expect the output voltage V_{coil} to increase with R_D at a high number of windings N, since a higher number of windings N requires a higher R_D. However, R_o also increases with an increasing number of windings N (see Fig. 3.10) and thus the ratio of the voltage divider decreases, as shown in Fig. 3.18. From this point of view, a Rogowski coil with a low number of windings and a large width $w_{winding}$ is preferred to achieve a high

Fig. 3.18 Ratio of the voltage divider of a planar Rogowski coil for different number of windings N and winding widths $w_{winding}$

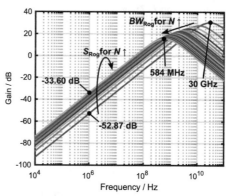

Fig. 3.19 Transfer behavior $R(s)$ of planar Rogowski coils for different number of windings N

ratio of the voltage divider between R_o and R_D and thus a high sensitivity S_{Rog} of the Rogowski coil.

The transfer behavior $R(s)$ or rather the sensitivity S_{Rog} of a Rogowski coil for the proposed area for an exemplary wire width $w_{winding} = 5\,\mu m$ and different number of windings N is shown in Fig. 3.19. The graph indicates a strong dependence of the resulting bandwidth BW_{Rog} and the sensitivity S_{Rog} on the number of windings N. The sensitivity S_{Rog} at a signal frequency of 1 MHz increases from $-52.87\,dB$ to $-33.60\,dB$, but at the same time the resulting bandwidth BW_{Rog} decreases from 30 GHz to 584 MHz for an increasing number of windings N.

Figure 3.20 shows the sensitivity S_{Rog} at a frequency of 1 MHz. The parasitic inductance L_o and capacitance C_o affect the transfer behavior in the high-frequency region, close to the natural frequency f_o. Therefore, the consideration at a frequency of 1 MHz allows to neglect these two parasitic components and S_{Rog} depends only on the mutual inductance M_p, the parasitic resistance R_o, and the damping resistance R_D. The sensitivity S_{Rog} increases due to the increasing mutual inductance M_p for an increasing number of windings N, as expected from Fig. 3.9. At high N, the drop in the ratio of the voltage divider R_o and R_D is greater than the rising value of

Fig. 3.20 Sensitivity S_{Rog} of a planar Rogowski coil for different number of windings N and winding widths $w_{winding}$ at a frequency of 1 MHz

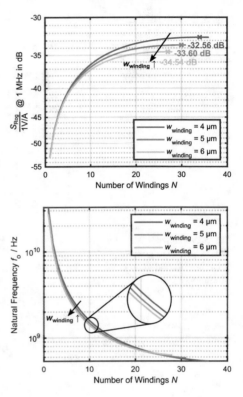

Fig. 3.21 Bandwidth BW_{Rog} of a planar Rogowski coil for different number of windings N and winding widths $w_{winding}$ and $D = 1$

M_p which gives S_{Rog} a maximum. Consequently, there is an explicit value of N to achieve the highest S_{Rog}. Regardless of N, the smaller the $w_{winding}$, the higher the S_{Rog}, see Fig. 3.20.

The resulting bandwidth BW_{Rog} or f_o of the Rogowski coil for $D = 1$ is shown in Fig. 3.21. As expected from the increasing inductance L_o and capacitance C_o, the dominant pole f_o of the Rogowski coil goes to lower frequencies for an increasing number of windings N. In addition, for a low number of windings N, a high BW_{Rog} can be achieved by a small winding width $w_{winding}$. Thus, a small $w_{winding}$ is recommended up to a certain number of windings to achieve the highest bandwidth BW_{Rog}.

Based on the previous comparisons, the following correlation can be derived for an optimized Rogowski coil design for on-chip current sensing:

1. High sensitivity S_{Rog} requires a large area for the coil. At the same time, it is important that the coil is as long as possible (parallel to the power line), see Fig. 3.8.

Fig. 3.22 Trade-off between sensitivity S_{Rog} and bandwidth BW_{Rog} of a planar Rogowski coil

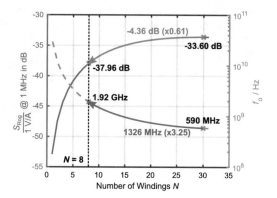

2. The winding width $w_{winding}$ must be as small as possible to achieve the highest sensitivity S_{Rog} (see Fig. 3.20) and highest bandwidth BW_{Rog} (see Fig. 3.21).
3. The space $s_{winding}$ between the windings also has to be minimal to achieve a high sensitivity S_{Rog}. At the same time the inductance L_0 increases, which can be neglected, fortunately.

The minimal width $w_{winding}$ and the minimal space $s_{winding}$ are given by the metal layer of the used technology. Therefore, for the final trade-off all geometric parameters (area, $w_{winding}$ and $s_{winding}$) are explicitly given. Figure 3.22 shows the resulting sensitivity S_{Rog} on left axis and the natural frequency f_0 on right axis for $D = 1$. The bandwidth BW_{Rog} is here equal to f_0. The highest sensitivity S_{Rog} is achieved for $N = 30$ windings with $f_0 = 590\,\text{MHz}$. The target overall bandwidth for on-chip current sensing is $50\text{–}100\,\text{MHz}$. In order to ensure that the coil does not limit the bandwidth or have any negative effects on the transfer behavior of the overall current sensing, this work proposes a bandwidth of $2\,\text{GHz}$ for the Rogowski coil itself. By the reduction to $N = 8$ windings, f_0 increases to $1.92\,\text{GHz}$, while there is only a small reduction in S_{Rog} ($-4.36\,\text{dB}$). In this case, the bandwidth BW_{Rog} is more critical than the sensitivity S_{Rog}, since the reduced sensitivity can be compensated by the subsequent sensor front-end of the Rogowski coil.

Fig. 3.23 Calculated and
simulated transfer behavior
$R(s)$ of a planar Rogowski
coil

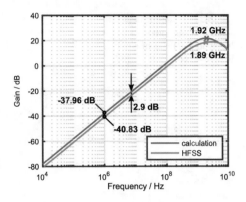

In conclusion, taking into account the parasitic components (R_o, L_o, and C_o) of the coil and the mutual inductance M_p between the power line and the coil, a bandwidth $BW_{Rog} = 1.92$ GHz and a sensitivity of -37.96 dB at a signal frequency of 1 MHz can be achieved with $N = 8$ windings for the proposed coil area of 700×500 µm.

3.2.4 Verification with 3D Field Simulation

The introduced coil design with $N = 8$ and the mentioned parameters of the geometry are verified by Ansys HFSS. Figure 3.23 shows the calculated and the simulated transfer behavior of the Rogowski coil with a damping resistor $R_D = 460\,\Omega$. The natural frequencies f_o of both graphs match very well and the calculated value $f_o = 1.92$ GHz by the induced analytical method was confirmed by HFSS.

The simulated transfer behavior is slightly lower (-2.9 dB) than the calculated. This results from the different mutual inductances M_p, the calculated inductance of 2.12 nH is higher than the simulated inductance of 1.52 nH. The difference occurs from the simplification of the coil geometry in Sect. 3.2.1 and the current distribution in the power line and the resulting dispersion of the magnetic field. The used approach in Eq. 3.10 with GMD g considering different widths of the power line and windings, but assumes, the current I_{meas} flows through an infinitesimal thin line with the distance of g. In the proposed design $w_{powerline} \gg w_{winding}$ applies and the distance between the power line and the coil is very small. Therefore, the assumed magnetic field $\overrightarrow{B}$ in the calculation differs from the field simulation and results in a small offset of the transfer behavior. Figure 3.24 shows the cross-section of the Rogowski coil and the magnetic field $\overrightarrow{B}$ for a signal current $I_{meas} = 1$ A. The magnetic field decreases with the distance to the power line, but is in the calculation higher than in the simulation.

In the HFSS simulation results, the parasitic components of the coil can be extracted considering the admittance Y_o of the coil. While the resistance R_o

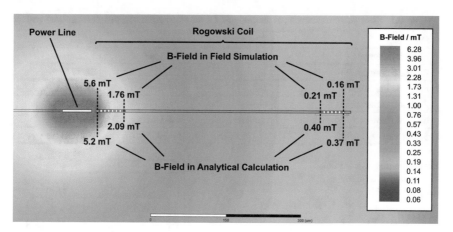

Fig. 3.24 Cross-section of a planar Rogowski coil with magnetic field $\vec{B}$

Fig. 3.25 Extracted inductance L_o by HFSS

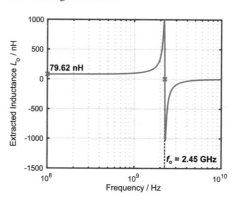

corresponds to the real path of Y_o at low frequencies with

$$R_o = \Re \left(\frac{1}{Y_o(f)} \right), \tag{3.21}$$

the inductance L_o depends on the imaginary path of Y_o and can be calculated by

$$L_o = \frac{1}{2\pi f} \cdot \Im \left(\frac{1}{Y_o(f)} \right). \tag{3.22}$$

Figure 3.25 shows the extracted inductance L_o by HFSS. The extracted $L_o = 79.62\,\text{nH}$ matches with the calculated value $L_o = 79.46\,\text{nH}$. At $L_o = 0$ the natural frequency f_o can be determined at 2.45 GHz. This value is higher than expected from the transfer behavior in Fig. 3.23, since here the influence of the resistance R_o and R_D are not considered (see Eq. 3.4). The parasitic capacitance $C_o = 101.92\,\text{fF}$

Table 3.1 Comparison between calculation and HFSS simulation of a planar Rogowski coil

Parameter	Symbol	Calculation	HFSS
Mutual inductance	M_p	2.12 nH	1.52 nH
Parasitic resistance	R_o	23.95 Ω	22.77 Ω
Parasitic inductance	L_o	79.46 nH	79.62 nH
Parasitic capacitance	C_o	91.40 fF	101.92 fF
Natural frequency/bandwidth	f_o/BW_{Rog}	1.92 GHz	1.89 GHz
Sensitivity @ 1 MHz	$S_{Rog}/(1V/A)$	−37.96 dB	−40.83 dB

Fig. 3.26 Equivalent circuit of an integrated planar Rogowski coil

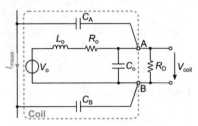

can be calculated with the extracted values of the coil R_o and L_o and R_D, by transforming Eq. 3.4.

A comparison between the analytical calculation and the field simulation is shown in Table 3.1. The introduced analytical calculation of the Rogowski coil matches well with the extracted components by HFSS field simulation. Therefore, the presented method in Sect. 3.2.3 allows to achieve the best trade-off between the bandwidth BW_{Rog} and the sensitivity S_{Rog} of the Rogowski coil.

3.2.5 Layout

For integrated on-chip current sensing with a planar Rogowski coil, the conventional equivalent circuit must be extended by the parasitic capacitance C_A and C_B between the power line and the terminals A and B of the coil, as shown in Fig. 3.26.

Because of the asymmetrical layout, the parasitic capacitances of the coil are not equal ($C_A \neq C_B$) and the coupling to terminal A and B is therefore unbalanced. Thus, a voltage swing of the power line produces a voltage swing at the output of the coil, which is in the range of the output signal for current sensing. Especially at low frequencies and low currents, the output signal of the Rogowski coil can be less than 100 μV. For example, the Rogowski coil, introduced in Sect. 3.2.2, generates an output signal with an amplitude of 91 μV for a sinusoidal current I_{meas} with an amplitude of 100 mA at a frequency of 100 kHz.

Fig. 3.27 Capacitive coupling into a planar Rogowski coil for a voltage swing of 50 V with a slope of 8 V/ns

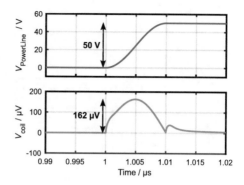

A transient voltage swing of the power line with an amplitude of 50 V and a maximum slope of 8 V/ns is shown in Fig. 3.27 ($I_{meas} = 0\,A$). The generated voltage swing at the differential output of the coil of 162 μV cannot be distinguished from a real current sensing signal. Such a transition occurs in fast switching voltage converters with integrated MOSFET switches and high input voltages [23]. For voltage converters with lower input voltage as presented in [24], slopes up to 80 V/ns can occur.

To optimize the parasitic coupling between the power line and the coil, the layout itself must be considered. Figure 3.28 shows a symmetrical layout of the coil, proposed as part of this work, which results in equal capacitances C_A and C_B. Hence, any voltage swing of the power line appears as a common-mode signal, which cancels in the subsequent sensing stage.

The modified layout changes the trade-off between coupling and bandwidth. For a fair comparison between the conventional and optimized layout, the design parameters and number of windings are kept equal, only the symmetry is changing as indicated in Fig. 3.28. The improved layout of the coil results in slightly lower coupling (-0.86 dB) and a lower bandwidth as shown in Fig. 3.29. But the voltage swing at the output of the coil can be reduced to 1.6 μV, which is an improvement by a factor 100, as shown in Fig. 3.30.

In conclusion, the layout of a planar Rogowski coil has large influence to the parasitic coupling between the power line and the output of the coil. Only a symmetrical layout ($C_A = C_B$) allows on-chip current sensing in switching applications. Due to the equal parasitic capacitance C_A and C_B only common-mode signals occur at the output of the coil for voltage swings of the power line. The proposed symmetrical planar Rogowski coil was implemented in a 180 nm HV CMOS technology, see Fig. 3.31. The optimized coil is placed besides a power line, which carries the current to be measured, at a distance of 10 μm. Therefore, galvanically isolated current sensing with superior characteristics is possible in integrated switching applications.

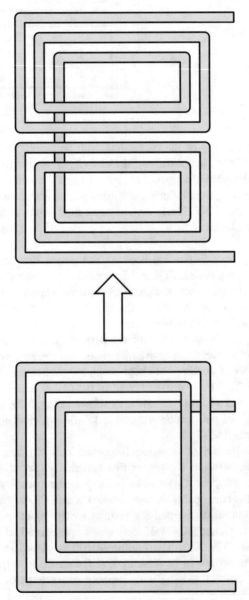

Fig. 3.28 Improved symmetrical layout of an integrated planar Rogowski coil for switching applications

Fig. 3.29 HFSS simulated
transfer behavior $R(s)$ of a
conventional and improved
symmetrical planar Rogowski
coil

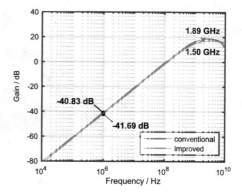

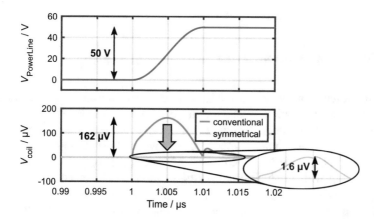

Fig. 3.30 Improvement of capacitive coupling into a symmetrical planar Rogowski coil for a
voltage swing of 50 V with a slope of 8 V/ns

Fig. 3.31 Photograph of the
implemented symmetrical
planar Rogowski coil in upper
metal

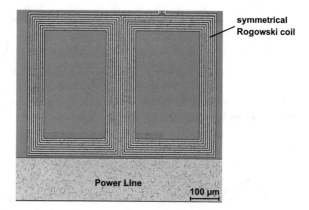

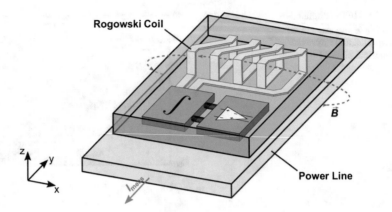

Fig. 3.32 Off-chip current sensing with an integrated helix-shaped Rogowski coil

3.3 Rogowski Coil for Off-Chip Current Sensing

Off-chip current sensing is essential for high-power applications to provide the required galvanic isolation as derived in Sect. 2.1.2. A bandwidth of more than 10 MHz is required to measure current changes up to 3 kA/µs with a high accuracy and with a short delay. In this work, the first known helix-shaped Rogowski coil on IC level is implemented for off-chip current sensing [2]. To get a high sensitivity S_{Rog}, the chip with the helix-shaped Rogowski coil is intended to be mounted on top of a power module or a PCB and measures the magnetic field $\vec{B}$ in x-direction, as shown in Fig. 3.32. Some publications use planar coils for off-chip current sensing, but they have high demands on positioning and require two large coils for an acceptable sensitivity [25, 26].

The distance between the Rogowski coil and the power line is minimized by back-grinded chips with total thickness of 60 µm instead of the regular thickness of 250 µm. This improves the sensitivity by a factor of two, since the magnetic field $\vec{B}$ is inversely proportional to the distance to the power line. By considering the parasitic components of the coil a bandwidth of 200 MHz is achieved.

There are many degrees of freedom for the design of the helix-shaped coil. The area determines the cross-section of the coil and maximum number and size of windings, as for the planar coil. Furthermore, for the design of the helix-shaped coil, the cross-section also depends on the height and number of available metal layers of the used technology. The available area of the coil is the first value to be defined. In the following considerations, this is set to a maximum size of 1,000 µm × 750 µm as an example. All comparisons for the design of the helix-shaped coil in following sections refer to this area.

Section 3.3.1 explores the coupling and thus the sensitivity of a helix-shaped Rogowski coil for off-chip current sensing. In Sect. 3.3.2 the relevant parasitic components are calculated and in Sect. 3.3.3 the resulting trade-offs are discussed.

3.3.1   Mutual Inductance

The mutual inductance M represents the magnetic coupling between the power line and the Rogowski coil, as mentioned for the on-chip current sensing in Sect. 3.2.1. While, for the helix-shaped Rogowski coil the thickness t_{chip} of the chip is also important and has to be minimal. The smaller the t_{chip}, the smaller the distance between the power line and the coil and therefore the higher the mutual inductance M_v. The mutual inductance M_v between the power line and the helix-shaped coil is defined as

$$M_v = \frac{\mu_0 \cdot w_{coil} \cdot N}{2\pi} \cdot \ln\left(\frac{g_{outer}}{g_{inner}}\right). \tag{3.23}$$

M_v is proportional to the width w_{coil} of the coil and the number of windings N and depends on the GMD between the power line and coil, similar to the planar Rogowski coil. To achieve a high mutual inductance, g_{inner} and therefore the thickness of the chip must be minimal and the height h_{coil} must be maximal to get a large cross-section area A_{coil} of the coil, as shown in Fig. 3.33a. The GMD g_{inner} and g_{outer} of the helix-shaped Rogowski coil can be calculated similar to the planar coil. While in Eq. 3.11, the width of the power line and the width of the winding of the coil is replaced by the thickness of the power line and thickness of the used metal layer in the chip, respectively.

For the implementation of the coil, the bottom metal layer and the top metal layer are used to achieve the maximal height h_{coil}, as shown in Fig. 3.33a. Hence, this is limited by the IC technology. Most relevant technologies reach maximum metal layer distance in the range of 8 µm. Figure 3.33b shows the top view of a simplified helix-shaped Rogowski coil layout. For a given area, the number of possible windings N depends on the width $w_{winding}$ of the coil and the space $s_{winding}$ between. The smaller these dimensions, the higher the possible number of windings and the higher the mutual inductance M_v.

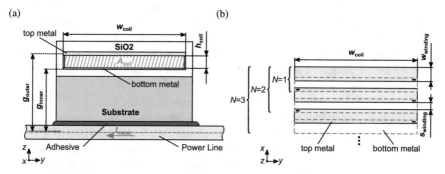

Fig. 3.33 Helix-shaped Rogowski coil: (**a**) cross-section of the coil in a chip directly mounted on a PCB power line and (**b**) top view of the helix-shaped coil

Table 3.2 Measured sensitivity S_{Rog} for different thicknesses t_{chip} of the test chip

	$S_{Rog}/(1\text{V/A})$ @ 100 kHz	$S_{Rog}/(1\text{V/A})$ @ 1 MHz
$t_{chip} = 60\,\mu m$	−50.25 dB	−49.91 dB
$t_{chip} = 250\,\mu m$	−56.08 dB	−55.50 dB
Difference	5.83 dB	5.59 dB

Fig. 3.34 Mutual inductances M_v of a helix-shaped Rogowski coil for different number of windings N and winding widths $w_{winding}$ with $s_{winding} = 5\,\mu m$

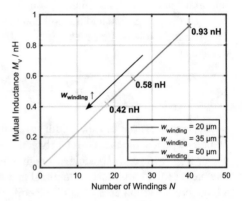

In the following the improvement of the coupling for different chip thicknesses is considered. For a fair comparison between the chip with regular thickness of 250 μm and the grinded chip with a thickness of 60 μm, each chip has the same coil design and sensor front-end. Table 3.2 shows the measured improvement for different thicknesses and signal frequencies. The grinded chip achieves a doubling (approximately 6 dB) of sensitivity compared to the regular chip thickness. This improvement is independent of the coil design and is directly related to the higher mutual inductance M_v as the distance between the power line and the coil is reduced. For this reason, a thickness $t_{chip} = 60\,\mu m$ is preferred and used for the following calculations.

Figure 3.34 shows the resulting mutual inductance M_v for different number of windings N and winding width $w_{winding}$. M_v is proportional to N as expected from Eq. 3.23. With an increasing winding width $w_{winding}$ the maximum number of possible windings decreases for the given area, whereby the maximum M_v also decreases. In general, M_v is significantly lower compared to the mutual inductance of the planar Rogowski coil for on-chip current sensing in Sect. 3.2.1. This results from the larger distance between the power line and the coil and the much smaller cross-section A_{coil} of the coil.

In conclusion, the mutual inductance M_v depends on the thickness of the chip and is proportional to the number of windings N. Due to a minimal thickness $t_{chip} = 60\,\mu m$ of the chip, M_v can be doubled compared to the regular thickness of 250 μm. To further increase M_v, the wire width $w_{winding}$ and consequently the space $s_{winding}$ between must be minimal to get the highest number of possible windings N for a given area. Generally, as larger the area A_{coil} of the coil as higher M_v. Similar to the planar implementation, the influence of increasing parasitic components, such as the resistance R_0, inductance L_0, and capacitances C_0, will impact the transfer behavior of the Rogowski coil. A large area A_{coil} and a high number of windings N reduce sensitivity S_{Rog} and bandwidth BW_{Rog} due to these parasitic components.

3.3.2   Parasitic Components

The parasitic components of the coil, namely resistance R_0, inductance L_0, and capacitances C_0, have a strong influence on the transfer behavior on the Rogowski coil. These parasitic elements must be as small as possible to achieve the highest performance of the Rogowski coil. As in the planar implementation, L_0 and C_0 will limit the bandwidth BW_{Rog} and R_0 will degrade the sensitivity S_{Rog} of the sensor. Due to the multilayer design of the helix-shaped coil especially R_0 and C_0 are much larger. The parasitic components are investigated in the following for different coil designs and for the given range ($1000 \times 750\,\mu m$).

Parasitic Resistance R_0

The parasitic resistance R_0 forms a voltage divider with the required damping resistance R_D and must be as small as possible, as mentioned for on-chip current sensing in Sect. 3.2.2. For a helix-shaped Rogowski coil, R_0 is proportional to the number of windings N. The resistance $R_{0,N}$ of one winding is the sum of the resistance of the bottom and top metal layer and of the via stacks and is defined as

$$R_{0,N} = \left(R_{S,bottom} + R_{S,top}\right) \frac{w_{coil}}{w_{winding}} + 2R_{via}. \qquad (3.24)$$

The resistance of the used metal layers depends on the corresponding sheet resistance R_S and the dimensions of the winding. The total resistance R_0 for different winding widths $w_{winding}$ is shown in Fig. 3.35. R_0 is proportional to the number of windings N and decreases with an increasing $w_{winding}$.

For a high sensitivity S_{Rog} of the coil, R_0 has to be as low as possible. To achieve a low R_0 the winding width $w_{winding}$ must be large and the number of windings N low, but at the same time the mutual inductance M_v decreases as a result. Therefore, there is a design trade-off between the resistance R_0 and the mutual inductance M_v for the given coil area, as will be discussed in Sect. 3.3.3.

Parasitic Inductance L_0

Similar to the planar Rogowski coil, the inductance L_0 limits the bandwidth BW_{Rog} of the coil and must be as low as possible. L_0 of the helix-shaped coil consists

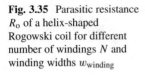

Fig. 3.35 Parasitic resistance R_0 of a helix-shaped Rogowski coil for different number of windings N and winding widths $w_{winding}$

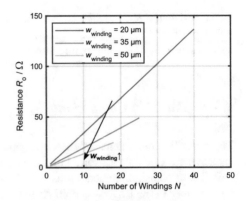

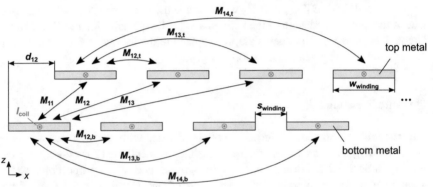

Fig. 3.36 Dimensions for the calculation of mutual inductance M_+ or M_- between windings of the helix-shaped Rogowski coil

of the same parts as the planar coil (self-inductance and positive/negative mutual inductance) and can be calculated according Eq. 3.14. While for the self-inductances L_S both metal layers (bottom and top) must be considered. For the calculation of the sum of the positive and negative mutual inductances M_+ and M_- all parallel segments of the coil must be considered. Whereby M_- only occurs between the bottom and top metal layer, as shown in Fig. 3.36.

The positive mutual inductance of two segments can be calculated similar to the planar implementation with the Greenhouse formula Eq. 3.17, since the relevant segments are in the same layer and have the same length. For the calculation of the negative mutual inductance, there are several approaches for integrated helix-shaped coils. While Hoer [27] and Grover [28] use the exact dimensions of the segments, the Greenhouse method, as introduced for the planar coil calculation in Sect. 3.2.2, uses the GMD between the segments. As part of this work, these three methods have been implemented and compared, the difference between them is < 2 %. Since the calculation by Hoer results in the lowest M_- and therefore in the highest inductance L_0 (worst case) for small dimensions of the coil, this method is

Fig. 3.37 Proportion of the parasitic inductance L_o of a helix-shaped Rogowski coil for different number of windings N

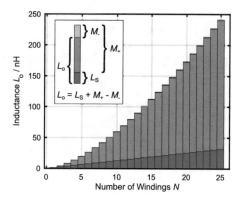

used for the following calculations. Hence, the negative mutual inductance between two segments is defined as

$$M_- = 0.001 \left[x \ln \left(x + \sqrt{x^2 + h_{coil}^2} \right) \right.$$
$$\left. - \sqrt{x^2 + h_{coil}^2} \right]_{d_{x,y}, \, d_{x,y}}^{d_{x,y} - w_{winding}, \, d_{x,y} + w_{winding}} (x). \qquad (3.25)$$

In Fig. 3.37 the partition of the inductance L_o for different numbers of windings N with $w_{winding} = 35\,\mu m$ is shown. Similar to the planar coil implementation, the most dominant part of L_o is the positive mutual inductance M_+ between the windings. This proportion is lowered by M_-, where this can be neglected in this coil design, as shown in Fig. 3.37.

The dominant mutual inductance M_+ can be reduced by an increasing width $w_{winding}$ of the windings or an increasing space $s_{winding}$ between the windings. Both result in an increasing distance between the centers of the windings, see Fig. 3.36. The increasing inductance L_o for different winding widths $w_{winding}$ and an increasing number of windings N is shown in Fig. 3.38a. With an increasing $w_{winding}$ the inductance L_o can be significantly reduced. This results due to the increasing distance between the centers of segments and the decreasing self-inductance L_S. Figure 3.38b shows L_o for an increasing space $s_{winding}$ between the windings. This will also reduce L_o, but there is only an improvement of the positive mutual inductance M_+. The geometric change of dimensions in Fig. 3.38a is equal to Fig. 3.38b. This results in the same distances between the centers of the segments and same number of possible windings. The optimization of $w_{winding}$ is slightly more sufficient as $s_{winding}$ and a lower L_o can be achieved with the same geometric change, e.g., $L_o = 239\,nH$ can be achieved when $w_{winding}$ is increased by $15\,\mu m$, while only $L_o = 242\,nH$ can be reached if $s_{winding}$ is increased by $15\,\mu m$.

Generally, the inductance L_o must be minimal to achieve a high bandwidth BW_{Rog} of the Rogowski coil. The most dominant part of L_o is the positive mutual inductance M_+ between the segments of the coil. M_+ can be reduced

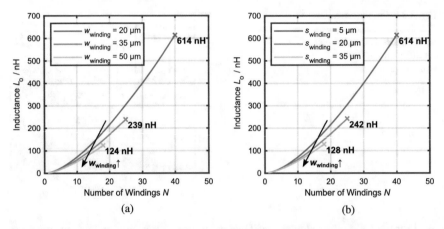

Fig. 3.38 Inductance L_o of a helix-shaped Rogowski coil for different number of windings N and **(a)** different winding widths $w_{winding}$ and **(b)** different spaces $s_{winding}$ between

by increasing the winding width $w_{winding}$ or space $s_{winding}$ between. While an increasing $w_{winding}$ is slightly more sufficient since the self-inductance L_S will also be reduced compared to an increasing $s_{winding}$. Additionally, a reduced number of winding N results in a lower L_o and thus a higher bandwidth BW_{Rog} of the coil. Therefore, there is a trade-off between the mutual inductance M_v and L_o. The number of windings N and the dimensions of the coil should be large to achieve a high sensitivity (high M_v) but at the same time the bandwidth BW_{Rog} decreases due to the increasing L_o.

Parasitic Capacitance C_o

Similar to the planar Rogowski coil, the parasitic capacitance C_o limits the bandwidth BW_{Rog} of the coil and must be as low as possible. For a helix-shaped coil, the substrate capacitance C_{sub} mainly depends on the dimensions of the segments of the bottom layer and is defined as

$$C_{sub} = N \cdot \epsilon_{ox} \cdot \frac{w_{coil} \cdot w_{winding}}{t_{ox}}. \tag{3.26}$$

In the equivalent circuit of an integrated coil, C_{sub} is split into half and connected between the terminals of the coil and the substrate, see Appendix [21, 22]. For the calculation of C_o, these two capacitances are connected in series, therefore C_{sub} is multiplied by a factor of 0.25 in Eq. 3.27.

$$C_o = 0.25 \cdot C_{sub} \tag{3.27}$$

Fig. 3.39 Parasitic capacitance C_0 of a helix-shaped Rogowski coil for different number of windings N and winding widths $w_{winding}$

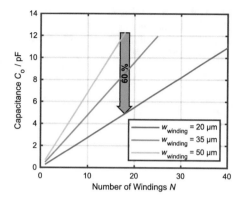

Figure 3.39 shows C_0 for different winding widths $w_{winding}$ and an increasing number of windings N. As expected from Eq. 3.27, C_0 increases with the number of windings N and the winding width $w_{winding}$.

Similar to the inductance L_0, an increasing C_0 will lower the bandwidth BW_{Rog} of the Rogowski coil. As with L_0, the number of windings N must be small to achieve a high bandwidth. But for C_0 also the width $w_{winding}$ has to be minimal, e.g., for the same number of windings C_0 can be improved by 60 % if $w_{winding}$ is reduced from 50 to 20 μm.

3.3.3 Design Trade-Offs

There are several trade-offs between the mutual inductance M_v and the parasitic components of helix-shaped the Rogowski coil, as discussed in Sect. 3.3.2. The transfer behavior $R(s)$ of the Rogowski coil for the proposed area for an exemplary space $s_{winding} = 20$ μm and different winding widths $w_{winding}$, respectively, number of windings N is shown in Fig. 3.40. The graph indicates a strong dependence on $w_{winding}$, respectively, N, for the resulting bandwidth BW_{Rog} and sensitivity S_{Rog}. These correlations are now compared to each other to achieve a high bandwidth BW_{Rog} with a high sensitivity S_{Rog} at the same time.

In order to achieve the best transfer behavior with $D = 1$ (see Sect. 3.1), the damping resistor R_D must be adapted for each geometric change of the Rogowski coil and can be calculated according to Eq. 3.6. Figure 3.41 shows a strong dependence on R_D for an increasing number of windings N and winding width $w_{winding}$. R_D is in the range between 24 Ω and 77 Ω and thus enables an integrated implementation.

In the following, the voltage divider between resistance R_0 and damping resistance R_D will be considered. A high ratio of the voltage divider and thus a high sensitivity of the Rogowski coil is necessary. Similar to the planar implementation in Sect. 3.2.3, one would expect that the output voltage V_{coil} over R_D increases for a

Fig. 3.40 Transfer behavior $R(s)$ of helix-shaped Rogowski coils for different number of windings N

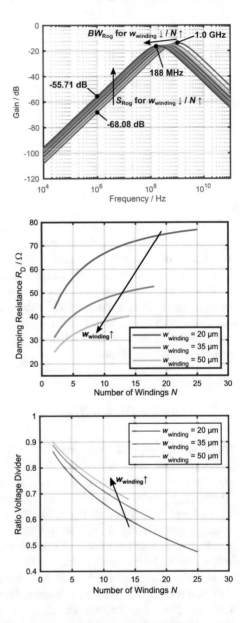

Fig. 3.41 Damping resistor R_D of a helix-shaped Rogowski coil for different number of windings N and winding widths $w_{winding}$

Fig. 3.42 Ratio of the voltage divider of a helix-shaped Rogowski coil for different number of windings N and winding widths $w_{winding}$

high number of windings N, since R_D increases for a higher number of windings in Fig. 3.41. However, as for the planar Rogowski coil, here the parasitic resistance R_o also increases for an increasing N (see Fig. 3.35) and thus the ratio of the voltage divider decreases, as shown in Fig. 3.42. From this point of view, a Rogowski coil with a low N is preferred to achieve a high ratio of the voltage divider between R_o and R_D.

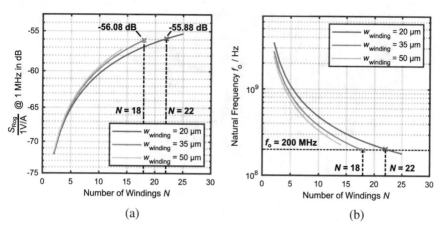

Fig. 3.43 Trade-off between (**a**) sensitivity S_{Rog} and (**b**) bandwidth BW_{Rog} of a helix-shaped Rogowski coil for different number of windings N and winding widths w_{winding} with $D = 1$

The parasitic inductance L_0 and capacitance C_0 affect the transfer behavior in the high-frequency region, close to the natural frequency f_0. Therefore, considering the sensitivity S_{Rog} at a frequency of 1 MHz allows to neglect these two parasitic components and S_{Rog} depends only on the mutual inductance M_v and the voltage divider between R_0 and R_D. In Fig. 3.43a, S_{Rog} increases for an increasing number of windings N due to the increasing mutual inductance M_v, as expected from Fig. 3.34. At high N, the drop in the ratio of the voltage divider R_0 and R_D is greater than the rising value of M_v whereby S_{Rog} saturates.

The resulting bandwidth BW_{Rog} of the Rogowski coil for $D = 1$ is shown in Fig. 3.43b. As expected from the increasing inductance L_0 and capacitance C_0, the natural frequency f_0 of the Rogowski coil goes to lower frequencies for an increasing number of windings N. With a small winding width w_{winding}, the capacitance C_0 is significantly reduced and a higher f_0, respectively, BW_{Rog} can be achieved. The target bandwidth for off-chip current sensing is > 10 MHz. In order to ensure that the coil does not limit the bandwidth or have any negative effects on the transfer behavior of the overall current sensing, this work proposes a bandwidth of 200 MHz for the Rogowski coil itself. This can be achieved with different w_{winding}, as shown in Fig. 3.43b. Considering these two coil designs ($w_{\text{winding}} = 20\,\mu\text{m}$ with $N = 22$ and $w_{\text{winding}} = 35\,\mu\text{m}$ with $N = 18$), approximately the same sensitivity S_{Rog} can be reached, as indicated in Fig. 3.43a. Therefore, there is a balance between w_{winding} and N of the Rogowski coil, which results in approximately the same sensitivity S_{Rog} for the same bandwidth BW_{Rog}.

Based on the previous comparisons, the following correlation can be derived for the design of a helix-shaped Rogowski coil:

1. The mutual inductance and thus the induced voltage in the coil is proportional to the available area of the coil, see Eq. 3.23.

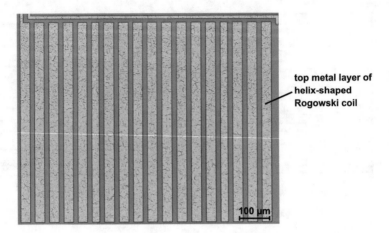

Fig. 3.44 Photograph of the implemented helix-shaped Rogowski coil for off-chip current sensing

2. Furthermore, the mutual inductance can be increased by back-grinded chips, see Table 3.2.
3. The space s_{winding} between the windings may not be minimal to achieve a low inductance L_0 and thus a high bandwidth, see Fig. 3.38b. In this work $s_{\text{winding}} = 20\,\mu\text{m}$ is chosen.
4. There is a balance between the bandwidth BW_{Rog} and the sensitivity S_{Rog}. Different designs of the Rogowski coil reaches the same BW_{Rog} and approximately the same S_{Rog} (see Fig. 3.43).

A photograph of the implemented helix-shaped Rogowski coil in a 180 nm HV CMOS technology with $N = 18$, $s_{\text{winding}} = 20\,\mu\text{m}$ and $w_{\text{winding}} = 35\,\mu\text{m}$ is shown in Fig. 3.44.

3.4   Summary

In this book the operation principle of the Rogowski coil is used for wide-bandwidth current sensing on IC level. Taking into account the mutual inductance M and the parasitic components of the coil such as the resistance R_0, inductance L_0, and capacitance C_0, a planar coil was optimized for on-chip current sensing and for the first time a helix-shaped coil was utilized and optimized for off-chip current sensing with high bandwidth and high sensitivity at the same time. To ensure that the coil does not limit the bandwidth and has no negative effect on the overall transfer behavior for current sensing, the bandwidth of the coil itself was chosen to be 20 times higher than that actually required for current sensing.

Fig. 3.45 Simplified
equivalent circuit of an
integrated coil

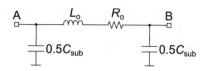

In order to achieve a high mutual inductance M and therefore a high sensitivity S_{Rog} of the Rogowski coil, the area of the coil must be as large as possible, the number of windings high and the distance between the power line and the coil minimal. But from this point of view, the influence of increasing parasitic components, such as resistance R_o, inductance L_o, and capacitance C_o, is neglected. These components must be considered for the transfer behavior of the Rogowski coil, since they will limit the bandwidth BW_{Rog} and reduce the sensitivity S_{Rog}.

With an optimized design of a symmetrical planar Rogowski coil for a given coil area of $700 \times 500\,\mu m$ for on-chip current sensing, a bandwidth BW_{Rog} of 1.5 GHz and a sensitivity S_{Rog} of -41.69 dB at 1 MHz is achieved. The symmetrical design reduces the differential capacitive voltage coupling between the power line and the coil by a factor of 100 compared to a conventional coil and enables current sensing with superior characteristics. The coil is placed besides a power line that carries the current to be measured with a distance of $10\,\mu m$. Therefore, a galvanically isolation for integrated current sensing of high-voltage switching applications is provided. The performance of the planar Rogowski coil is confirmed by HFSS 3D field simulations with very good matching.

For off-chip current sensing, the optimized helix-shaped Rogowski coil achieves a bandwidth of 200 MHz with a sensitivity of -55.88 dB at 1 MHz for an area of $1000 \times 750\,\mu m$. The distance between the Rogowski coil and the power line is optimized by back-grinded chips with a total thickness of $60\,\mu m$ in comparison to the regular thickness of $250\,\mu m$. This improves sensitivity by a factor of two. The implementation of a helix-shaped coil enables to place the chip on top of a current carrying conductor and to measure contactless the current flowing under it.

Appendix

Equivalent Circuit of an Integrated Coil

Figure 3.45 shows the simplified equivalent circuit on a fully integrated coil. The inductance L_o is the sum of the self-inductance of the wire and the mutual inductances between the windings of the coil, while R_o is the parasitic resistance of the coil. The capacitance C_{sub} represents the substrate capacitance between the metal layer of the coil and the substrate. This is divided into half and connected to terminals A and B, respectively.

In the equivalent circuit diagram of the Rogowski coil these two capacitances are connected in series whereby the capacitance between terminal A and B can be calculated with $0.25 C_{sub}$.

References

1. Funk, T., & Wicht, B. (2018). A Fully integrated DC to 75 MHz current sensing circuit with on-chip Rogowski coil. In *Proceedings of the IEEE Custom Integrated Circuits Conference (CICC)* (pp. 1–4). https://doi.org/10.1109/CICC.2018.8357028.
2. Funk, T., Groeger, J., & Wicht, B. (2019). An integrated and galvanically isolated DC-to-15.3 MHz hybrid current sensor. In *Proceedings of the IEEE Applied Power Electronics Conference and Exposition (APEC)* (pp. 1010–1013).
3. Xiang, M., Gao, H., Zhao, B., Wang, C., & Tian, C. (2011). Analysis on transfer characteristics of Rogowski coil transducer to travelling wave. In *Proceedings of the International Conference on Advanced Power System Automation and Protection* (Vol. 2, pp. 1056–1059). https://doi.org/10.1109/APAP.2011.6180705.
4. Guillod, T., Gerber, D., Biela, J., & Muesing, A. (2012). Design of a PCB Rogowski coil based on the PEEC method. In *Proceedings of the Seventh International Conference on Integrated Power Electronics Systems (CIPS)* (pp. 1–6).
5. Wang, B., Wang, D., & Wu, W. (2009). A Rogowski coil current transducer designed for wide bandwidth current pulse measurement. In *Proceedings of the IEEE Sixth International Power Electronics and Motion Control Conference* (pp. 1246–1249). https://doi.org/10.1109/IPEMC.2009.5157575.
6. Nibir, S. J., Hauer, S., Biglarbegian, M., & Parkhideh, B. (2018). Wideband contactless current sensing using hybrid magnetoresistor-Rogowski sensor in high frequency power electronic converters. In *Proceedings of the IEEE Applied Power Electronics Conf. and Exposition (APEC)* (pp. 904–908). https://doi.org/10.1109/APEC.2018.8341121.
7. Dalessandro, L., Karrer, N., & Kolar, J. W. (2007). High-performance planar isolated current sensor for power electronics applications. *IEEE Transactions on Power Electronics, 22*(5), 1682–1692. ISSN: 0885-8993. https://doi.org/10.1109/TPEL.2007.904198.
8. Ho, G. K.Y., Fang, Y., Pong, B. M. H., & Hui, R. S. Y. (2017). Printed circuit board planar current transformer for GaN active diode. In *Proceedings of the IEEE Applied Power Electronics Conference and Exposition (APEC)* (pp. 2549–2553). https://doi.org/10.1109/APEC.2017.7931056.
9. Luo, Z., & Wei, X. (2016). Mutual inductance analysis of planar coils with misalignment for wireless power transfer systems in electric vehicle. In *Proceedings of the IEEE Vehicle Power and Propulsion Conference (VPPC)* (pp. 1–6). https://doi.org/10.1109/VPPC.2016.7791733.
10. Peters, C., & Manoli, Y. (2007). Improved and accelerated analytical calculation algorithm for multi-wire coils to power wireless sensor systems. In *Proceedings of the Actuators and Microsystems Conference on TRANSDUCERS 2007—2007 International Solid-State Sensors* (pp. 1927–1930). https://doi.org/10.1109/SENSOR.2007.4300536.
11. Qing, C., Hong-bin, L., Ming-ming, Z., & Yan-bin, L. (2006). Design and characteristics of two Rogowski coils based on printed circuit board. *IEEE Transactions on Instrumentation and Measurement, 55*(3), 939–943. ISSN: 0018-9456. https://doi.org/10.1109/TIM.2006.873788.
12. Wang, K., Yang, X., Li, H., Wang, L., & Jain, P. (2018). A high-bandwidth integrated current measurement for detecting switching current of fast GaN devices. *IEEE Transactions on Power Electronics, 33*(7), 6199–6210. ISSN: 0885-8993. https://doi.org/10.1109/TPEL.2017.2749249.
13. Rezaee, M., & Heydari, H. (2008). Mutual inductances comparison in Rogowski coil with circular and rectangular cross-sections and its improvement. In *Proceedings of the Third IEEE*

Conference on Industrial Electronics and Applications (pp. 1507–1511). https://doi.org/10.1109/ICIEA.2008.4582770.

14. Shafiq, M., Hussain, G. A., Kütt, L., & Lehtonen, M. (2014). Effect of geometrical parameters on high frequency performance of Rogowski coil for partial discharge measurements. *Measurement, 49*, 126–137. ISSN: 0263-2241. https://doi.org/10.1016/j.measurement.2013.11.048. http://www.sciencedirect.com/science/article/pii/S0263224113006015.

15. Rosa, E. B., & Grover, F. W. (1916). *Formulas and tables for the calculation of mutual and self-inductance*. Washington: United States Government Printing Office. https://catalog.hathitrust.org/Record/100163561.

16. Higgins, T. J. (1947). Theory and application of complex logarithmic and geometric mean distances. *Transactions of the American Institute of Electrical Engineers, 66*(1), 12–16. ISSN: 0096-3860. https://doi.org/10.1109/T-AIEE.1947.5059399.

17. Greenhouse, H. (1974). Design of planar rectangular microelectronic inductors. *IEEE Transactions on Parts, Hybrids, and Packaging, 10*(2), 101–109. ISSN: 0361-1000. https://doi.org/10.1109/TPHP.1974.1134841

18. Peters, C., & Manoli, Y. (2008). Inductance calculation of planar multi-layer and multi-wire coils: An analytical approach. *Sensors and Actuators A: Physical, 145–146*, 394–404. ISSN: 0924-4247. https://doi.org/doi.org/10.1016/j.sna.2007.11.003. http://www.sciencedirect.com/science/article/pii/S0924424707008485.

19. Duan, Z., Guo, Y. X., & Kwong, D. L. (2012). Rectangular coils optimization for wireless power transmission. *Radio Science, 47*(3), 1–10. ISSN: 1944-799X. https://doi.org/10.1029/2011RS004970.

20. Kazimierczuk, M. K. (2014). *High-frequency magnetic components* (2nd ed, p. 757). Description based upon print version of record. Chichester: Wiley. Online-Resource. ISBN: 9781118717738. http://gbv.eblib.com/patron/FullRecord.aspx?p=1511094.

21. Passos, F. M. (2013). *Modeling of integrated inductors for RF circuit design*. PhD thesis. Faculdade de Ciências e Tecnologia Universidade Nova de Lisboa.

22. Yue, C. P., & Wong, S. S. (2000). Physical modeling of spiral inductors on silicon. *IEEE Transactions on Electron Devices, 47*(3), 560–568. ISSN: 0018-9383. https://doi.org/10.1109/16.824729.

23. Wittmann, J., Rosahl, T., & Wicht, B. (2014). A 50 V high-speed level shifter with high dv/dt immunity for multi-MHz DCDC converters. In *Proceedings of the ESSCIRC 2014—40th European Solid State Circuits Conference (ESSCIRC)* (pp. 151–154). https://doi.org/10.1109/ESSCIRC.2014.6942044.

24. Wittmann, J., Barner, A., Rosahl, T., & Wicht, B. (2016). An 18 V input 10 MHz Buck converter with 125 ps mixed-signal dead time control. *IEEE Journal of Solid-State Circuits, 51*(7), 1705–1715. ISSN: 0018-9200. https://doi.org/10.1109/JSSC.2016.2550498.

25. Jiang, J., & Makinwa, K. (2016). A hybrid multi-path CMOS magnetic sensor with 76 ppm/°C sensitivity drift. In *Proceedings of the ESSCIRC Conference 2016: 42nd European Solid-State Circuits Conference* (pp. 397–400). https://doi.org/10.1109/ESSCIRC.2016.7598325.

26. Jiang, J., & Makinwa, K. A. A. (2017). Multipath wide-bandwidth CMOS magnetic sensors. *IEEE Journal of Solid-State Circuits, 52*(1), 198–209. ISSN: 0018-9200. https://doi.org/10.1109/JSSC.2016.2619711.

27. Hoer, C., & Love, C. (1965). Exact inductance equations for rectangular conductors with applications to more complicated geometries. *Journal of Research of the National Bureau of Standards C. Engineering and Instrumentation, 69*(2), 127–137.

28. Grover, F. W. (2009). *Inductance calculations*. Dover Books on Electrical Engineering. New York: Dover Publications. ISBN: 9780486318356.

Chapter 4
Rogowksi Coil Sensor Front-End

The output signal V_{coil} of a Rogowski coil needs to be processed by a sensor front-end to get a constant transfer behavior in the frequency domain and a high sensitivity for the subsequent signal evaluation, as indicated in Fig. 4.1. V_{coil} is proportional to the signal frequency f_{meas} and has a phase shift of $+90°$ to the signal current I_{meas} due to the differentiating transfer behavior of the Rogowski coil itself. Both dependencies can be compensated with an integrating behavior of the sensor front-end such that the output voltage $V_{out,R}$ reflects the waveform of I_{meas}.

Figure 4.2 shows conceptually ① the ideal compensation of the differentiating transfer behavior of the Rogowski coil in the frequency domain. Thereby, a constant gain can be achieved from $f_{Rog,min}$ up to the natural frequency f_o of the coil itself. In order to cover a wide frequency range with the Rogowski path, $f_{Rog,min}$ must be low. To compensate for the resulting low sensitivity, the front-end requires an amplification ②. This amplifies the signal by A_o and achieves an acceptable sensitivity $S_{out,R}$ of the Rogowski path.

The output voltage V_{coil} of a Rogowski coil has a wide amplitude range. The number of decades in the frequency domain in which the current is to be measured corresponds to the number of decades in the amplitude domain. Additionally, the number of decades in the amplitude domain of the signal current I_{meas} corresponds also to the number of the decades of V_{coil}. For example, to cover a frequency range of three decades and a current amplitude range of two decades, the amplitude of the output voltage V_{coil} has a range of five decades. The minimal frequency $f_{Rog,min}$ of the Rogowski path and lower limit of the signal current I_{meas} is limited by induced noise, respectively, the DC gain A_o of the sensor front-end. The upper limit of the frequency range and I_{meas} is defined by the bandwidth and the maximal input voltage range of the sensor front-end.

Several integrating sensor front-end concepts for Rogowski coils were published for discrete electronics and also on PCB. For large signal currents I_{meas} (in the range of several kA) no amplification by the front-end is required and only a low-

T. Funk, B. Wicht, *Integrated Wide-Bandwidth Current Sensing*,
https://doi.org/10.1007/978-3-030-53250-5_4

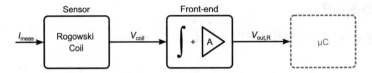

Fig. 4.1 Signal processing for current sensing with a Rogowski coil

Fig. 4.2 Compensation of differentiating transfer behavior of a Rogowski coil in the frequency domain

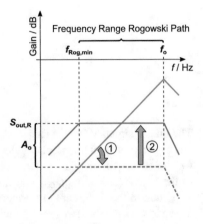

resistance damping resistor can be connected to the output of the coil [1]. Most state-of-the-art concepts use a single-ended active integrator circuit to increase the sensitivity for low signal currents I_{meas} and low signal frequencies f_{meas} and cover a wide-bandwidth under consideration of the connecting line between the coil and the amplifier to [2–6]. However, this does not suppress the capacitive coupling between the power line and the coil and is only for discrete implementations recommended. The concept for integrated coils, reported in [7], consists of a fully differential implementation to achieve a high sensitivity and a high Common-mode rejection ratio (CMRR). However, the maximum bandwidth of this current sensor is limited to 3 MHz.

An active sensor front-end for open-loop current sensing is proposed in Sect. 4.1 [8, 9]. The integrated two-stage integrator circuit enables chopping below the signal bandwidth, which results in a significant improvement in noise and thus an extension of the sensing range to lower frequencies and lower signal currents. This enables wide-bandwidth current sensing for on-chip and off-chip current sensing.

A compensated active sensor front-end for closed-loop current sensing is proposed in Sect. 4.2. The output signal of the Rogowski coil is compensated by DC and AC feedback, thus a constant amplitude sensing range to higher frequencies is achieved. Furthermore, a significant improvement of the sensitivity is achieved with a comparable bandwidth.

4.1 Open-Loop Sensing

This book proposes an active integrator sensor front-end for open-loop wide-bandwidth current sensing from 15 kHz up to 75 MHz for integrated Rogowski coils. It comprises a two-stage integrator with chopping, resulting in 2.2 mVrms output noise.

Figure 4.3a shows the architecture of the proposed sensor front-end with two integrator stages. Wide-bandwidth sensing is enabled by the combination of the chopped Integrator 1 for the low-frequency components and Integrator 2 for the high-frequency components of the signal current I_{meas}, as shown in Fig. 4.3b [8]. To achieve a constant gain of the overall transfer behavior of the Rogowski path, the dominant pole f_{p2} of Integrator 2 needs to match the zero f_{z1} of Integrator 1, i.e. $f_{p2} = f_{z1} = 1.3\,\text{MHz}$. Therefore, f_{p2} can be trimmed by the bias current $I_{bias,A2}$ of amplifier A_2. The minimal frequency $f_{Rog,min}$ for current sensing with the Rogowski path is defined by the dominant pole f_{p1} of Integrator 1, since here $f_{Rog,min} = f_{p1}$.

Due to the combination of two active integrator stages, the requirements for high amplification of the individual amplifier decrease, but both amplifiers require a high bandwidth. The 2nd order pole of the amplifier A_2 must even be above the desired

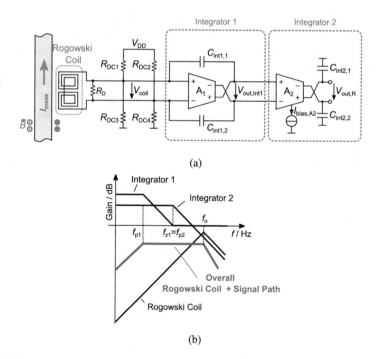

(a)

(b)

Fig. 4.3 Sensor front-end for open-loop current sensing including **(a)** simplified block diagram and **(b)** transfer characteristic

sensing bandwidth of $> 50\,\text{MHz}$, as otherwise the output of the Rogowski coil is overcompensated.

4.1.1   First Integrator Stage

Integrator 1 consists of the chopped amplifier A_1, implemented according to Appendix, and the feedback capacitances $C_{\text{int}1,1}$ and $C_{\text{int}1,2}$. Due to $C_{\text{int}1}$ and consideration of the output resistance $r_{\text{out},A1}$ of A_1, this integrator stage has a second order transfer function, which is defined as

$$\frac{V_{\text{out,Int1}}(s)}{V_{\text{coil}}(s)} = \frac{C_{\text{int}1}\, r_{\text{out},A1}\, s^2 + C_{\text{int}1}\, r_{\text{out},A1}\, \omega_{A1}\, s + A_{o,A1}\, \omega_{A1}}{C_{\text{int}1}\, r_{\text{out},A1}\, s^2 + (C_{\text{int}1}\, r_{\text{out},A1}\, \omega_{A1} + 1)\, s + \omega_{A1}}. \tag{4.1}$$

While $A_{o,A1}$ is the DC gain and ω_{A1} is the bandwidth of the amplifier A_1. The resulting dominant pole

$$f_{p1} = \frac{1}{2\pi\, C_{\text{int}1} r_{\text{out},A1}} \tag{4.2}$$

defines the minimal frequency $f_{\text{Rog,min}}$ for current sensing with the Rogowski path ($f_{\text{Rog,min}} = f_{p1}$). This allows to set $f_{\text{Rog,min}}$ by the feedback capacitance $C_{\text{int},1}$ and the output resistance $r_{\text{out},A1}$. The zero f_{z1} results from the design of the amplifier A_1 and the chosen feedback capacitances $C_{\text{int}1,1}$ and $C_{\text{int}1,2}$. Since $C_{\text{int}1,1}$ and $C_{\text{int}1,2}$ are connected between the positive input and output, respectively, negative input and output of amplifier A_1, the Integrator 1 has a phase shift of $+90°$ at zero f_{z1}, as shown in Fig. 4.4. This enables two-stage integration with a correct phase response.

Table 4.1 shows the simulated parameters of Integrator 1 across technology corners and temperatures from $-40\,°\text{C}$ to $150\,°\text{C}$. The DC gain $A_{o,A1}$ varies only slightly. The variation of the dominant pole f_{p1} and thus the variation of minimal

Fig. 4.4 Nominal transfer behavior of Integrator 1 for open-loop current sensing

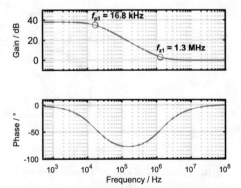

Table 4.1 Transfer behavior of Integrator 1 including corner and temperature variation for open-loop current sensing with $C_{int1,1} = C_{int1,2} = 462\,pF$

Parameter	Symbol	Nominal	Variation
DC gain	$A_{o,A1}$	38.1 dB	36.1– 39.1 dB
Dominant pole	f_{p1}	16.8 kHz	12.8–21.7 kHz
Zero	f_{z1}	1.3 MHz	0.8–1.9 MHz

frequency $f_{Rog,min}$ of the Rogowski path cannot be neglected. But this can be set by the bias current of amplifier A_1. The resulting change of the zero f_{z1} is then compensated by the bias current of the Integrator 2. Thus, the variation of the parameters in Table 4.1 are not critical and can be compensated by the corresponding bias currents.

4.1.2 Second Integrator Stage

To expand the bandwidth of the Rogowski path towards higher frequencies, the dominant pole f_{p2} of Integrator 2 needs to match with the resulting zero f_{z1} of Integrator 1 ($f_{p2} = f_{z1}$). Thus the $+20\,dB$/decade rise of the Rogowski coil is also compensated for high-frequency components $f_{meas} > f_{z1}$ and a constant gain of the overall transfer behavior can be achieved (see Fig. 4.3b).

High-frequency components of the signal current I_{meas} cause large amplitudes of the output voltage V_{coil} of the Rogowski coil due to the differentiating transfer behavior of itself. Since Integrator 2 is used to integrate high-frequency components, the differential input of the proposed amplifier A_2 is cascoded by M_{C1} and M_{C2}, as shown in Fig. 4.5. Therefore, the capacitive coupling to $V_{M1/2}$ by $C_{par1/2}$ is blocked and a high bandwidth for current sensing is achieved [8]. Without M_{C1} and M_{C2}, the maximum bandwidth for current sensing would be limited to $<20\,MHz$, for a maximum signal current of 1 A.

The pole f_{p2} of Integrator 2 is set by the integrator capacitances $C_{int2,1}$ and $C_{int2,2}$, see Fig. 4.3a, which can be adjusted by the bias current $I_{bias,A2}$ to match with the zero f_{z1} of Integrator 1. To cover a wide frequency range by the Rogowski path, the 2nd pole $f_{p2,2nd}$ of the amplifier A_2 must be at high frequencies, since the bandwidth $f_{Rog,max}$ of the Rogowski path is thereby limited ($f_{Rog,max} = f_{p2,2nd}$). The pole $f_{p2,2nd}$ causes a decrease of $-40\,dB$ of A_2 in the frequency domain and leads to an overcompensation of the output signal of the Rogowski coil. Therefore, the output stage of A_2 is optimized for low parasitic capacitances. For this purpose, the output transistors M_6 to M_{11} are kept small and the source-follower M_{F1} and M_{F2} buffers the output of A_2 for the required Common-mode feedback (CMFB). Due to the smaller sizing of the output transistors (with constant W/L), the output resistance $r_{out,A2}$ and thus the gain $A_{o,A2}$ decreases compared to A_1. However, this shifts $f_{p2,2nd}$ into the high-frequency range and allows wide-bandwidth current sensing.

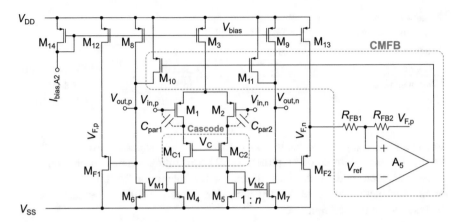

Fig. 4.5 Schematic of cascoded amplifier A_2 of Integrator 2 for open-loop current sensing

Table 4.2 Transfer behavior of Integrator 2 for a bias current $I_{bias,A2} = 38\,\mu A$ including corner and temperature variation for open-loop current sensing

Parameter	Symbol	Nominal	Variation
DC gain	$A_{o,A2}$	26.9 dB	26.0– 27.6 dB
Dominant pole	f_{p2}	1.3 MHz	1.0–1.6 MHz
Second pole	$f_{p2,2nd}$	96.0 MHz	79.0–110.7 MHz

Table 4.2 shows the parameters of Integrator 2 for a bias current $I_{bias,A2} = 38\,\mu A$ across technology corners and temperatures from $-40\,°C$ to $150\,°C$. For nominal simulation, the dominant pole f_{p2} matches the zero f_{z1} of Integrator 1. The variation of f_{p2} can be compensated by the bias current of the amplifier A_2 to ensure a constant gain of the overall transfer behavior. The nominal bandwidth $f_{Rog,max}$ of the Rogowski path is limited to 96.0 MHz by the 2nd pole $f_{p2,2nd}$ of the amplifier A_2. This 2nd pole has a large variation, but is in the desired frequency range ($>50\,MHz$) for on-chip current sensing.

Figure 4.6 shows the measured signal transfer behavior of the proposed two-stage integrator circuit with a planar Rogowski coil and a helix-shaped Rogowski coil [8, 9]. The measured transfer behavior represents the sensitivity $S_{out,R} = V_{out,R}/I_{meas}$ and frequency range ($f_{Rog,min}$ to $f_{Rog,max}$) of both Rogowski paths. For on-chip current sensing with a planar Rogowski coil, sinusoidal signal currents I_{meas} with an amplitude of 0.5 A are generated by an RF broadband power amplifier (6 kHz–110 MHz). Frequency components $15\,kHz < f_{meas} < 75\,MHz$ are measured with a sensitivity $S_{out,R}$ of $-27.2\,dB$ related to I_{meas} and a variation $< 5\,\%$ ($43.7\,mV/A \pm 2.2\,mV/A$). The measured bandwidth is slightly lower than expected from the simulation results in Table 4.2. This is caused by the lower bias currents of amplifiers A_1 and A_2. These currents must be lower compared to the simulation in order to measure the desired minimum signal frequencies of 15 kHz.

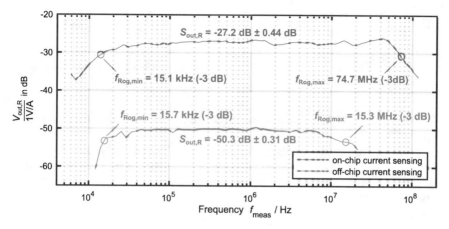

Fig. 4.6 Measured transfer behavior of on-chip current sensing with a planar Rogowski coil and off-chip current sensing with a helix-shaped Rogowski coil

Because of the lower sensitivity of the helix-shaped Rogowski coil for off-chip current sensing, sinusoidal signal currents I_{meas} with an amplitude of 1.4 A are generated by an RF broadband power amplifier. Currents between 16 kHz and 15 MHz can be measured with a sensitivity $S_{out,R}$ of -50.3 dB related to I_{meas}, which corresponds to 3.1 mV/A. Because of the increasing capacitive coupling between the power line and the chip at high frequencies of the signal current I_{meas}, the bandwidth for off-chip current sensing is lower than the measured bandwidth for on-chip current sensing. Nevertheless, both measured transfer behaviors confirm that the bandwidth $f_{Rog,max}$ of the front-end can be extended by the 2nd integrator stage and the required bandwidth for on-chip current sensing (> 50 MHz) and off-chip current sensing (>10 MHz) is achieved. Thus, these current sensors exceed the state of the art significantly, as will be discussed in Sect. 4.4.

4.1.3   Low-Frequency Error Cancelation

At low frequencies, noise and offset are the most dominant errors sources of the implemented CMOS amplifiers and must be reduced below the minimum signal amplitude of the Rogowski coil. For the resolution of the sensor, noise represents a critical parameter, since the noise determines the minimum sensing signal that is still distinguishable from the noise and that can therefore be measured with the sensor. The noise voltage $V_{out,R,noise}$ of the sensor front-end must therefore be as low as possible. The offset voltage V_{OS} of differential input pairs can be as large as 10 mV [10] and will be amplified by each stage in a multi-stage amplifier circuit, whereby the output of the sensor front-end $V_{out,R}$ would saturate to the supply rails.

Fig. 4.7 Simplified profile of
the PSD of a sensor front-end

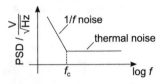

Therefore, especially the offset of the first amplifier must be small (in the range of
a few μV).

Noise in integrated circuits consists mainly of thermal noise and $1/f$ noise.
Thermal noise is caused by thermal energy that creates a random movement of
charge carriers in a conductive material and has a constant Power spectral density
(PSD) [11, 12]. $1/f$ noise is found in all active devices and is mainly caused by
the defects in the interface between the gate oxide and the silicon substrate. It
is inversely proportional to the frequency and the dominant error source at low
frequencies [12, 13]. The resulting input related PSD of a sensor front-end is
indicated in Fig. 4.7. The cut-off frequency f_c defines the transition in which the
$1/f$ noise or thermal noise dominates.

Low-frequency error cancelation of Integrator 1 is required, since otherwise an
rms noise voltage $V_{out,R,noise}$ of 27.9 mV results at the output of the sensor front-
end. Additionally, the input offset voltage V_{OS} of A_1 is 3.4 mV (3σ) whereby the
output of the sensor front-end $V_{out,R}$ would saturate to the supply rails. Therefore,
a cancelation technique, such as auto-zeroing or chopping is needed to counteract
these problems [13].

Auto-Zeroing is a time-discrete method based on sampling and compensation of
low-frequency noise and offset voltage V_{OS}. These low-frequency disturbing signals
are stored on a capacitance in one switching phase and subtracted at the output in the
second switching phase. Basically, there are three different auto-zeroing topologies:
input offset storage, output offset storage, and a closed-loop cancelation with an
compensation amplifier [13]. The basic topology of the auto-zeroing topology with
a compensation amplifier is shown in Fig. 4.8a. It overcomes the limitation in
bandwidth and gain of topologies with series capacitances at the input or output of
the amplifier. The topology with a compensation amplifier consists of the amplifier
$A_{1,1}$, to amplify the input signal V_{coil}, the capacitances C_1 and C_2, to store the low-
frequency interference signals, and the compensation amplifier $A_{1,2}$. During the
first clock phase ϕ_1 the topology is separated from the output V_{coil} of the Rogowski
coil and the output $V_{out,Int1}$ of Integrator 1. The input of amplifier $A_{1,1}$ is shorted
and the capacitances C_1 and C_2 are connected to the outputs of both amplifiers.
The DC offset voltages V_{OS1} and V_{OS2} as well as the low-frequency interference
signals of the two amplifiers are sampled during ϕ_1 on the capacitances C_1 and
C_2. In the second switching phase ϕ_2 the amplifier $A_{1,1}$ is connected to the output
of the Rogowski coil and amplifies this output signal V_{coil}. The amplifiers $A_{1,1/2}$
are both connected to the output $V_{out,Int1}$ of the Integrator 1. The low-frequency
interference signals, which are stored on the capacitors C_1 and C_2, are compensated
by the amplifier $A_{1,2}$. Due to the clocked system and the parasitic capacitances of

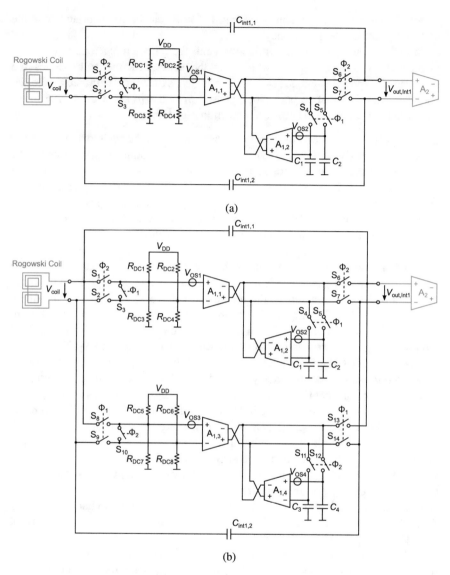

Fig. 4.8 Block diagram of auto-zeroing topology with (**a**) a compensation amplifier and (**b**) the extension of the ping-pong technique

the switches S_{4-7}, charge injections occurs, which is amplified by the amplifier $A_{1,2}$ and results in an error in the output voltage $V_{out,Int1}$. The transconductance g_m of $A_{1,2}$ should therefore typically be 50 times lower than the g_m of $A_{1,1}$ [13]. Dimensioning of C_1 and C_2 and switches S_4 and S_5 is a compromise between maximum auto-zeroing frequency f_{az} and magnitude of the voltage peaks caused by the charge injection. With auto-zeroing, $1/f$ noise can be compensated up to the half of the auto-zeroing frequency $f_{az}/2$.

In order to achieve a time-continuous gain, the auto-zeroing topology is supplemented by a second path, which has the same structure and enables the so-called ping-pong technique [14–16]. The switches of the two paths are controlled complementary, so that in both clock phases ϕ_1 and ϕ_2 one of the two paths is connected to the input and output, as shown in Fig. 4.8b.

Chopping is a continuous-time method based on the frequency modulation of the signal and noise, whereby the signal in the frequency domain is separated from the noise and the noise is canceled through a filter. The simplified block diagram of a conventional chopping topology is shown in Fig. 4.9a. The resulting transient signals are indicated in Fig. 4.9b and the frequency modulation principle is indicated in Fig. 4.9c. For the input signal V_{coil}, the restriction applies that V_{coil} may have a maximum bandwidth of $f_{ch}/2$ so that no aliasing effects due to sub-sampling occur [14].

The input signal V_{coil} is modulated by the first chopper element CH_1 to the chopper frequency f_{ch} and its odd harmonic, resulting in the voltage V_1 at the input of the amplifier A_1. In addition, the modulated voltage V_1 is shifted by the offset voltage V_{OS} of A_1. The amplifier A_1 amplifies this modulated signal, resulting in the voltage V_2 at the output of the amplifier. The voltage V_2 contains the amplified modulated signal as well as the noise and offset of the amplifier A_1. The second chopper element CH_2 demodulates the voltage V_2. The noise is simultaneously modulated to the chopper frequency f_{ch} and its odd harmonic. Voltage V_3 after the second chopper element CH_2 contains the amplified signal as well as the modulated noise. The residual ripple is caused by the offset of the amplifier A_1. Since the noise is separated from the low-frequency signal in the frequency domain, the noise can be filtered and the ripple can be reduced with the low-pass filter (LPF). The output voltage $V_{out,Int1}$ contains the amplified signal and the filtered noise, while the offset voltage V_{OS} affects $V_{out,Int1}$ only in the form of a voltage ripple. With the required chopper frequency f_{ch} in the MHz region, the noise can be significantly reduced, since especially the low-frequency $1/f$ noise can be filtered up to half the chopper frequency. However, the limitation is that f_{ch} must be much higher than the signal bandwidth $f_{Rog,max}$ [13].

Unlike conventional chopping, with the combination of two integrator stages, this work proposes to choose a chopper frequency f_{ch} below the signal bandwidth $f_{Rog,max}$. The chopper frequency f_{ch} must be higher than the zero f_{z1} to suppress in-band frequency components of f_{ch} [8]. Furthermore, f_{ch} has to be lower than the dominant pole $f_{o,A1}$ to guarantee high gain of A_1, as indicated in Fig. 4.10a. This leads to high demands on the signal bandwidth of A_1 ($f_{o,A1} \gg f_{z1}$), but allows f_{ch} below $f_{Rog,max}$. Therefore, the amplifier A_1 is optimized for high bandwidth according Appendix and achieves $f_{o,A1} = 26\,\text{MHz}$ in this design. Since Integrator 1 has feedback capacitances $C_{int1,1}$ and $C_{int1,2}$, the low-pass filter for ripple reduction and noise filtering after demodulation can be omitted, as shown in Fig. 4.10b. $C_{int1,1}$ and $C_{int1,2}$ cause the required filtering of the output signal $V_{out,Int1}$, since

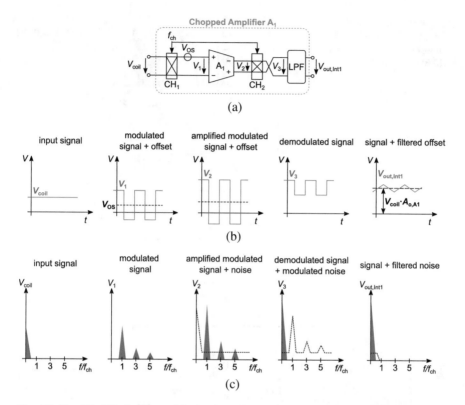

Fig. 4.9 Simplified block diagram (**a**) of a chopped amplifier with chopping principle in the time domain (**b**) and in the frequency domain (**c**)

the dominant pole, according Eq. 4.2, of this circuit is in the low kHz region (see Table 4.1). As an additional effect, the voltage peaks caused by charge injection of the chopper elements CH_1 and CH_2 are damped and the resulting error at the output $V_{out,Int1}$ of the Integrator 1 can be neglected.

Figure 4.11 shows the measured spectrum of the Rogowski path without a signal current for different chopper frequencies f_{ch} ($I_{meas} = 0\,A$). If f_{ch} is lower than the zero f_{z1} of Integrator 1 ($f_{z1} \approx 1.3\,MHz$), f_{ch} and its odds harmonics are present in the spectrum at the output voltage $V_{out,R}$ of the sensor front-end. For $f_{ch} > 5\,MHz$ the chopper frequency disappears in the noise floor. This confirms the previous consideration that the chopper frequency must be higher than f_{z1} but can be lower than the signal bandwidth.

The spectrum of the output voltage $V_{out,R}$ for off-chip current sensing is shown in Fig. 4.12. The chopper frequency f_{ch} is 10 MHz to obtain a small output ripple, and the signal current I_{meas} has a frequency of 8 MHz with an amplitude of 1 A. The expected increase at low frequencies ($f < 2\,MHz$) caused by $1/f$ noise is suppressed by chopping. By means of the introduced two-stage integrator, f_{ch}

Fig. 4.10 Chopping of
Integrator 1: (**a**) selection of
chopper frequency f_{ch} and
(**b**) block diagram

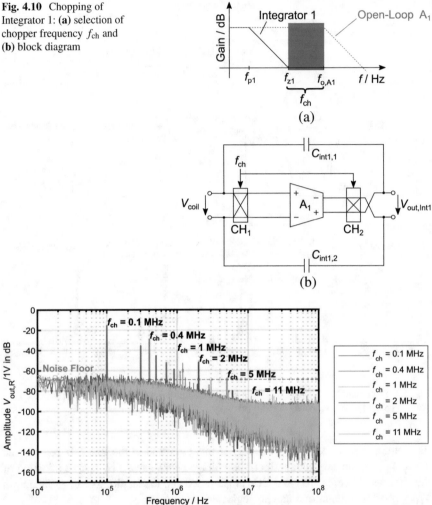

Fig. 4.11 Measured spectrum of the Rogowski path for different chopper frequencies f_{ch} with
signal current $I_{meas} = 0 \, A$

disappears in thermal noise and only the signal frequency f_{meas} of the signal current
I_{meas} and its harmonics are dominant.

Comparison of the introduced auto-zeroing and chopping technique in the two-
stage integrator circuit for open-loop current sensing is shown in Table 4.3. Both
techniques can reduce noise by up to 94%. However, the implementation of auto-
zeroing is more complex due to the required ping-pong technique for time-continues
current sensing and it requires a large amount of area. Chopping has the advantage
that due to frequency modulation the offset voltage V_{OS} of the amplifier A_1 appears

Fig. 4.12 Measured
spectrum of the output
voltage $V_{out,R}$ of the
Rogowski path with a
helix-shaped coil

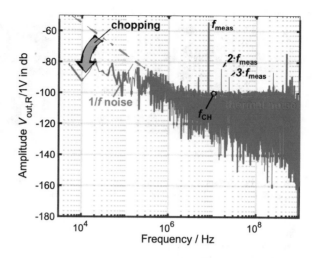

Table 4.3 Comparison
between auto-zeroing and
chopping in a two-stage
integrator circuit for
open-loop current sensing

	Auto-zeroing	Chopping
Noise reduction	+	+
Residual offset	+	++
Ripple	+	+
Area	−	++

only in the form of an output ripple. This voltage ripple is suppressed by the
large integrator capacitance C_{int1} and therefore does not require a low-pass filter
after demodulation, compared to conventional chopping. Additionally, due to the
configuration as a two-stage integrator circuit, a chopper frequency f_{ch} below signal
bandwidth can be chosen and overcomes the bandwidth limitation of a conventional
chopper architecture. This is the reason, why chopping is preferred over auto-
zeroing in this work.

4.1.4   Transient Current Sensing

Figure 4.13a shows a photograph of the implemented chip with Rogowski coils and
the sensor front-ends. To verify on-chip current sensing, the power line is connected
to external test circuits via bond wires. This allows to generate different signal
currents I_{meas} and to compare them with a reference measurement. For off-chip
current sensing, the chip with a total thickness of $60\,\mu m$ is mounted directly on
a PCB for contactless sensing of the signal current I_{meas} through the underlying
power line, as shown in Fig. 4.13b.

For the experimental evaluation of transient current sensing, signal currents I_{meas}
are generated by an RF broadband power amplifier. Figure 4.14 shows the transient
signals of the on-chip current sensing with a planar Rogowski coil for a 70 MHz

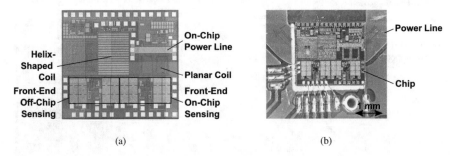

Fig. 4.13 Photograph of (**a**) the implemented chip including the front-ends, the planar and helix-shaped Rogowski coils and (**b**) the bare-chip mounted on top a PCB for off-chip current sensing

Fig. 4.14 Transient measurement result of on-chip current sensing at f_{meas}=70 MHz

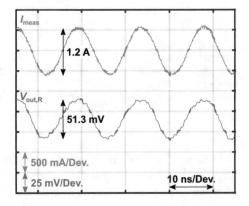

sinusoidal signal current I_{meas} with a peak-to-peak current of 1.2 A [8]. This high signal frequency was not achieved with prior art current sensors, as will discussed in Sect. 4.4. The sensing output $V_{\text{out,R}}$ has a peak-to-peak voltage of 51.3 mV, which corresponds to a sensitivity $S_{\text{out,R}}$ of 42.75 mV/A. The output voltage $V_{\text{out,R}}$ of the integrating sensor front-end reflects the waveform of the large signal current I_{meas}. The slightly lower sensitivity $S_{\text{out,R}}$ results from the limited bandwidth $f_{\text{Rog,max}}$ of the sensor front-end ($f_{\text{Rog,max}} = 75$ MHz).

The sensitivity for different amplitudes of I_{meas} for on-chip current sensing at a constant frequency f_{meas} is shown in Fig. 4.15. The output voltage $V_{\text{out,R}}$ increases linearly with the amplitude of I_{meas}, while the maximum peak-to-peak current is limited to 4.1 A by the RF broadband amplifier in the lab. Due to a chopper frequency $f_{\text{ch}} = 9$ MHz, the output noise voltage $V_{\text{out,R,noise}}$ can be reduced by a factor of 9.4 and minimum currents of 50 mArms can be measured. Without chopping the introduced implementation for on-chip current sensing design would be limited to 658 mArms of minimum detectable currents.

Figure 4.16 shows the measured step response of the on-chip current sensing for a pulsed signal current I_{meas} from 0 A to 1 A with a fast rise / fall-time of 10 ns.

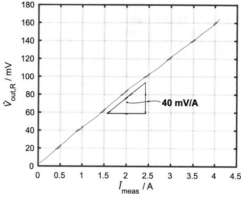

Fig. 4.15 Sensitivity measurement result of on-chip current sensing at $f_{meas} = 1\,\mathrm{MHz}$

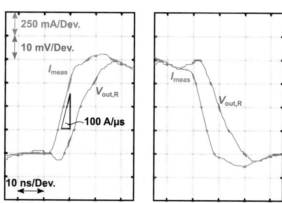

Fig. 4.16 Measured transient response of on-chip current sensing with a 1 A current pulse and a 100 A/μs rise time

The output voltage $V_{out,R}$ follows this pulse with a rise time of 10.6 ns at a delay of 6.7 ns [8].

The pulsed signal current I_{meas} for off-chip current sensing with a helix-shaped Rogowski coil is generated by the IGBT double-pulse setup with a regulated slew rate of 1 kA/μs. The current sensor is placed close to the emitter of the low-side IGBT to avoid parasitic coupling from the high-voltage nodes that may destroy it. Double-pulse measurements have been performed in order to generate the load current pulses of 60 A at a DC link voltage of 300 V. The proposed current sensor precisely resolves the full current transition, including non-idealities such as the reverse recovery peak, as shown in Fig. 4.17 [9].

In conclusion, with the presented open-loop sensor front-end, the required bandwidth for on-chip current sensing and off-chip current sensing is achieved. The two integrator stages allow chopping for low-frequency errors cancelation. A distinctive feature is that the chopper frequency f_{ch} can be selected lower than the signal bandwidth and disappears in the noise floor. The sensor front-end and the planar and helix-shaped Rogowski coils are implemented in a 180 nm HV CMOS technology.

Fig. 4.17 Measured transient response of off-chip current sensing with a 60 A current pulse and a 1 kA/µs rise time

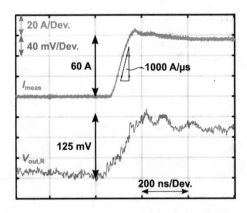

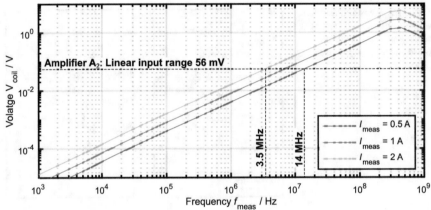

Fig. 4.18 Output voltage V_{coil} of a planar Rogowski coil for different signal amplitudes and frequencies of the signal current I_{meas}

4.2 Closed-Loop Sensing

The differentiating transfer behavior of the Rogowski coil results in an output voltage that is both proportional to the amplitude of the signal current I_{meas} and proportional to the signal frequency f_{meas}. Thus, the open-loop sensor front-end introduced in Sect. 4.1 has a strong limitation for large signal amplitudes at high signal frequencies. To simplify the analysis of this limitation, only the amplifier A_2 of Integrator 2 is considered in the following, since Integrator 1 is short circuited for high signal frequencies by the integration capacitors $C_{int1,1/2}$, see Fig. 4.3. The limitation for large amplitudes at high frequencies of I_{meas} results from the linear input range of A_2 and is defined in this work by the 1 dB compression point. Due to the saturation of A_2, the deviation of the output signal is −1 dB compared to the ideal amplified output signal. Figure 4.18 shows the output voltage V_{coil} of the planar Rogowski coil for different signal currents I_{meas} and the linear input range of

A_2 of 56 mV. Large signal amplitudes of I_{meas} already reach the linear gain range of A_2 at lower signal frequencies. As a result, the maximum linear current amplitude sensing range decreases towards high signal frequencies and only small amplitudes of I_{meas} can be measured at high signal frequencies.

In discrete implementations of the Rogowski coil and sensor front-end, this limitation of the amplitude sensing range has been solved by a passive integrator filter at the input of the sensor front-end, connected between the coil and the amplifier of the subsequent active integrator [6, 17]. The filter compensates for the differentiating transfer behavior of the Rogowski coil for high signal frequencies. Therefore, high signal amplitudes at high signal frequencies are damped at the input of the amplifier and linear amplification over a wide frequency range is ensured. Due to the variation of the absolute values of the passive components and the signal amplification, which is required to obtain sufficient sensitivity for current sensing, such a filter is not preferred for IC level implementation.

This book proposes the first sensor front-end for closed-loop current sensing for a Rogowski coil. Thanks to negative feedback, it inherently keeps the signal swing at the sensing input at a minimum. The implemented design achieves a wide frequency range from 17 kHz to 63 MHz. Due to a high-pass feedback, the differentiating transfer behavior of the Rogowski coil is compensated and it allows an extension of the amplitude sensing range at high frequencies. Furthermore, a high sensitivity of 447 mV/A and a low output offset of <1.5 mV is achieved. A particular challenge is that higher frequencies require a stronger signal compensation and the feedback requires a bandwidth of ≫100 MHz.

To measure high signal currents I_{meas} at high signal frequencies f_{meas}, a feedback $H(s)$ is added to the open-loop sensor front-end to compensate the output voltage of the Rogowski coil, as shown in Fig. 4.19. The induced voltage V_o is reduced by the compensation voltage V_{comp}. The effective voltage at the output of the coil V_{coil} is thus lower than the induced voltage V_o.

The transfer function of the closed-loop system is given as

$$\frac{V_{out,R}(s)}{I_{meas}(s)} = \frac{R(s)G(s)}{1 + G(s)H(s)}. \tag{4.3}$$

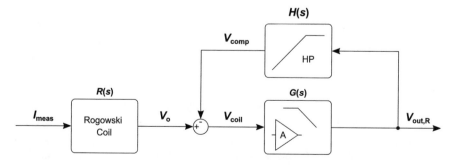

Fig. 4.19 Simplified block diagram of the sensor front-end for closed-loop sensing

If the loop gain is $G(s)H(s) \gg 1$, the transfer function can be simplified to $R(s)/H(s)$. Within the frequency range, in which this condition is fulfilled, the transfer function of the entire closed-loop system is thus defined by the Rogowski coil $R(s)$ and the feedback $H(s)$.

Due to the differentiating transfer behavior of the Rogowski coil, the amplitude of the induced voltage V_o gets larger at high frequencies. As a result, the feedback transfer function of $H(s)$ must be equivalent to a high-pass filter (with $+20\,dB$/decade gain increase).

The lower bandwidth limit $f_{Rog,min}$ and the upper bandwidth limit $f_{Rog,max}$ of the overall transfer function are defined by feedback $H(s)$ and loop gain $G(s)H(s)$. For frequencies above the cut-off frequency f_{hp} of the high-pass filter, the transfer function according to Eq. 4.4 results. The overall transfer function approaches the transfer function $R(s)$ of the Rogowski coil which increases with $+20\,dB$/decade.

$$\frac{V_{out,R}(s)}{I_{meas}(s)} \approx R(s) \qquad \text{with } G(s) \gg 1 \text{ and } H(s) \approx 1 \qquad (4.4)$$

Therefore, the upper bandwidth limit $f_{Rog,max}$ of the overall transfer function is defined by the cut-off frequency of the high-pass filter ($f_{Rog,max} = f_{hp}$). The lower limiting frequency $f_{Rog,min}$ is defined by the loop gain $G(s)H(s)$. In the frequency range below the dominant pole of $G(s)$, the gain of $G(s)$ is constant, while the feedback $H(s)$ increases with $+20\,dB$/decade to higher frequencies. In this case, the cut-off frequency $f_{Rog,min}$ is defined by $|G(s)||H(s)| = 1$.

4.2.1 Forward Path $G(s)$

From the shown transfer behavior for the closed-loop current sensing in Fig. 4.20 and the considerations of the upper and lower bandwidth $f_{Rog,max}$ and $f_{Rog,min}$, the relationship between the cut-off frequency f_{hp} and the DC gain of $G(s)$ is derived according to:

$$A_{o,G(s)} \geq (\log{(f_{Rog,max})} - \log{(f_{Rog,min})}) \cdot 20\,dB. \qquad (4.5)$$

Based on the bandwidth requirements between $50\,MHz$ and $100\,MHz$ for on-chip current sensing from Sect. 2.1.1 and the lower bandwidth limit for open-loop sensing, the DC gain of $G(s)$ can be determined. With $f_{Rog,min} = 15\,kHz$ and $f_{Rog,max} = 100\,MHz$, a DC gain of $A_{o,G(s)} \geq 76\,dB$ in required according to Eq. 4.5.

The forward path $G(s)$ is implemented with the two amplifier stages A_1 and A_2 according to Appendix, as shown in Fig. 4.21. In order to achieve a high gain over a wide frequency range, in this work, feedback capacitances C_{fb1} and C_{fb2} are applied to amplifier A_1. At high frequencies amplifier A_1 is shorted by these capacitors and Stage 1 has no negative contribution to the phase characteristic (positive phase

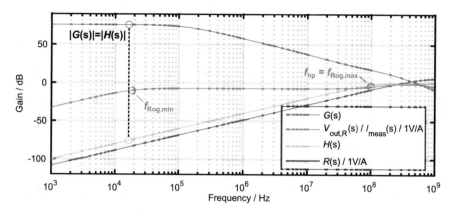

Fig. 4.20 Transfer behavior of closed-loop sensor front-end

Fig. 4.21 Forward path $G(s)$
of the sensor front-end for
closed-loop sensing

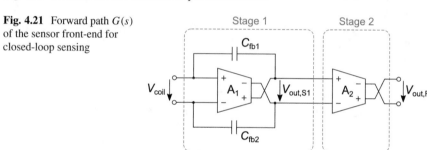

shift at the resulting zero, similar to open-loop sensing in Sect. 4.1). This allows to
shift the dominant pole of Stage 1 towards higher frequencies, which ensures the
required amplification in the high-frequency region. With $C_{fb1} = C_{fb2} = 8.1\,\text{pF}$
a phase margin of PM $> 23°$ at a transit frequency of 116 MHz (worst case across
technology corner and temperature from $-40\,°\text{C}$ to $150\,°\text{C}$) of the forward path $G(s)$
is achieved.

4.2.2 Feedback Considerations

In the target frequency range, the Rogowski coil can be considered in first order
approximation as a frequency-dependent voltage source with the internal resistance
$R_0 = 33\,\Omega$ (symmetrical planar Rogowski coil in Sect. 3.2.5). In order to achieve a
high attenuation of the induced voltage V_0 ($V_{coil} \to 0$), the feedback must represent
a short circuit ($R_L \to 0$). Thus the feedback must generate a maximum current
which corresponds to the short-circuit current of the coil, which is only limited by
the internal resistance R_0. For frequencies up to 100 MHz, the maximum short-
circuit current for the planar Rogowski coil is 6 mA, which must be compensated
by the output stage of the feedback $H(s)$. However, this large current requires

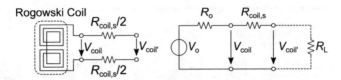

Fig. 4.22 Simplified equivalent circuit diagram of a Rogowski coil with additional series resistance $R_{coil,s}$ for closed-loop sensing

a low impedance output stage with large transistors, which makes the required bandwidth of 100 MHz difficult to achieve. To reduce the short-circuit current, this design proposes a resistor $R_{coil,s} = 3\,k\Omega$ in series to the Rogowski coil, as shown in Fig. 4.22. This lowers the maximum short-circuit current from 6 mA to 120 µA. Hence, the feedback does not require an amplifier with an output power stage (class A/AB).

As a result, the series resistor $R_{coil,s}$ together with Stage 1 of $G(s)$ forms a second order system which may be unstable. The transfer function with $R_{coil,s} \gg R_o$ can be calculated by

$$\frac{V_{out,S1}(s)}{V_{coil}(s)} = \frac{C_{fb}\, r_{out,A1}\, s^2 + C_{fb}\, r_{out,A1}\, \omega_{A1}\, s + A_{o,A1}\, \omega_{A1}}{b_1\, s^2 + b_2\, s + \omega_{A1}}$$

with

$$b_1 = (C_{fb}\, r_{out,A1} + C_{fb}\, R_{coil,s}/2)$$

$$b_2 = (C_{fb}\, \omega_{A1}\, R_{coil,s}/2 + C_{fb}\, r_{out,A1}\, \omega_{A1}$$
$$- A_{o,A1} \cdot C_{fb}\, \omega_{A1}\, R_{coil,s}/2 + 1). \tag{4.6}$$

For the system to remain stable, the resulting poles $s_{p1,2}$ must have a negative real part. These poles are defined as

$$s_{P1,2} = -\frac{C_{fb}\, \omega_{A1}(r_{out,A1} + (1 - A_{o,A1})R_{coil,s}/2) + 1}{k_2} \pm \frac{\sqrt{k_1}}{k_2} \tag{4.7}$$

with

$$k_1 = C_{fb}^2\, \omega_{A1}{}^2\, (R_{coil,s}/2 + r_{out,A1})^2 + (A_{o,A1}\, C_{fb}\, \omega_{A1}\, R_{coil,s}/2 - 1)^2$$
$$- 2\, C_{fb}\, \omega_{A1}\, (R_{coil,s}/2 + r_{out,A1})\, (A_{o,A1}\, C_{fb}\, \omega_{A1}\, R_{coil,s}/2 + 1) \tag{4.8}$$

$$k_2 = 2\, C_{fb}\, (r_{out,A1} + R_{coil,s}/2).$$

In order to ensure that both poles $s_{p1,2}$ have a negative real part, the first term of $s_{p1,2}$ must be negative. This results in the following stability criterion:

$$-\frac{C_{fb}\, \omega_{A1}(r_{out,A1} + (1 - A_{o,A1})R_{coil,s}/2) + 1}{k_2} < 0. \tag{4.9}$$

With $A_{o,A1} \gg 1$ and neglecting (+1), this criterion can be simplified to Eq. 4.10, which represents a useful design rule.

$$-\frac{C_{fb}\,\omega_{A1}}{k_2}\cdot(r_{out,A1} - A_{o,A1}\,R_{coil,s}/2) < 0 \quad \Longrightarrow \quad A_{o,A1}\,R_{coil,s}/2 < r_{out,A1}.$$

$$(4.10)$$

With the given output resistance of $r_{out,A1} \approx 16\,\text{k}\Omega$, a DC gain of $A_{o,A1} = 100$ and $R_{coil,s} = 3\,\text{k}\Omega$, this condition is not fulfilled and an unstable system results. For the stability of the circuit, the output resistance $r_{out,A1}$ must therefore be increased to

$$r_{out,A1} > A_{o,A1}\cdot\frac{R_{coil,s}}{2} = 150\,\text{k}\Omega. \qquad (4.11)$$

By changing the output resistance $r_{out,A1}$ of the existing amplifier, the DC gain $A_{o,A1}$ changes in first approximation by the same amount, whereby no improvement in stability is achieved. Thus, further changes or a complete redesign would be necessary. For this reason, this work proposes an additional resistor $R_{out,s}$ at the output of the existing amplifier A_1. It provides stability without changing the design of the existing amplifier, as shown in Fig. 4.23. The condition for stability according to Eq. 4.11 changes to

$$r_{out,A1} + R_{out,s} > A_{o,A1}\cdot\frac{R_{coil,s}}{2}. \qquad (4.12)$$

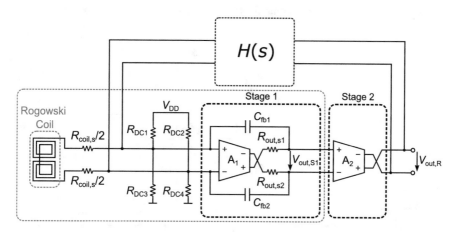

Fig. 4.23 Block diagram of the modified forward path for stability consideration of Stage 1 of the sensor front-end for closed-loop sensing

With the parameters stated above, stability of the system can be provided for $R_{out,s} > 134\,k\Omega$. The simulated poles $s_{p1,2}$ on transistor level for different values of $R_{out,s}$ are verified in Appendix. In order to guarantee stability with sufficient margin, the nominal resistance $R_{out,s}$ is increased to $170\,k\Omega$.

4.2.3   AC Signal Compensation

The previous consideration has shown that the differentiating transfer behavior of the Rogowski coil is compensated by the transfer behavior of a high-pass feedback $H(s)$. Figure 4.24 shows the implemented feedback $H(s)$. The high-pass filter consists of $R_{hp} = 4\,k\Omega$ and $C_{hp} = 398\,fF$ with the cut-off frequency $f_{hp} = 100\,MHz$. The voltage $V_{out,hp}$ at the output of the high-pass filter is buffered by the amplifier A_3 with a DC gain of approximately $0\,dB$ (set by $R_{hp,fb}$). A_3 with the resistors $R_{hp,fb}$ in feedback sees the Rogowski coil with the series resistor $R_{coil,s}$ as load.

The amplifier A_3 converts the voltage $V_{out,hp}$ into a proportional output current I_{comp}, which compensates the output signal of the Rogowski coil. For this purpose, the output stage of A_3 must be able to generate the optimized compensation current $I_{comp,max} = 120\,\mu A$, which results from the additional series resistance $R_{coil,s}$ in Fig. 4.22. The Gain-bandwidth (GBW) of amplifier A_3 must be higher than the cut-off frequency f_{hp} of the high-pass filter to buffer the voltage $V_{out,hp}$ up to f_{hp}. Furthermore, the dominant pole and the 2nd order pole of A_3 must be far from each other to achieve a positive phase margin for the overall loop gain $G(s)H(s)$.

In order to fulfill the demand for a high bandwidth, along with a low negative phase shift and sufficient driver strength at high frequencies, amplifier A_3 is implemented as a pseudo-differential symmetrical amplifier according to [18]. This

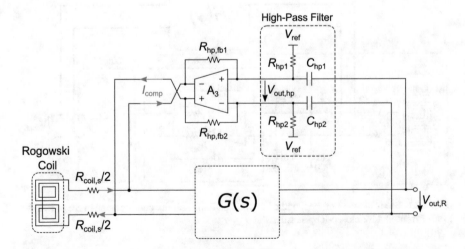

Fig. 4.24 Block diagram of high-pass feedback of the sensor front-end for closed-loop sensing

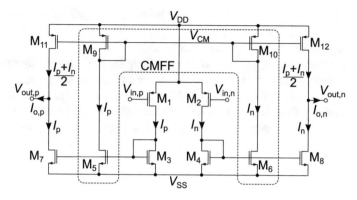

Fig. 4.25 Schematic of the pseudo-differential symmetrical amplifier with CMFF

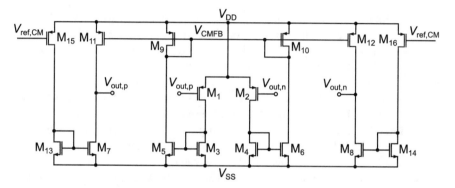

Fig. 4.26 CMFB stage of the pseudo-differential symmetrical amplifier

amplifier has only one high-impedance node (dominant pole) and the non-dominant pole is at very high frequencies. The amplifier consists of a pseudo-differential symmetrical stage in which the currents $I_{p/n}$ are mirrored to the output $V_{out,p}$ and $V_{out,n}$ by the current mirrors $M_3 = M_7$ and $M_4 = M_8$, as shown in Fig. 4.25. In addition to the architecture of the symmetrical amplifier, the Common-mode feedforward (CMFF) technique is implemented. The current mirrors $M_3 = M_5$ and $M_4 = M_6$ mirror the currents $I_{p/n}$ into a common-mode path and the differential signal components are canceled out. Thus, the common-mode voltage V_{CM} results, which is a measure for the common-mode voltage of the input signal. This implementation achieves a high CMRR (nominal simulation > 260 dB).

CMFF can suppress the input common-mode signals, but the DC operation point of the output $V_{out,p}$ and $V_{out,n}$ is not set [18]. Therefore a CMFB circuit is implemented in addition to the amplifier stage, Fig. 4.26.

The input stage of the CMFB is equal to the input stage of the pseudo-differential amplifier of Fig. 4.25. The inputs of the CMFB are connected to the outputs $V_{out,p}$ and $V_{out,n}$ of the amplifier stage. By canceling the differential input signals the voltage V_{CMFB} results, which depends on the common-mode output voltage of the amplifier. The voltage V_{CMFB} is compared by the current mirror $M_{13} = M_7$ and $M_{14} = M_8$ with the reference voltage $V_{ref,CM} = V_{DD}/2$, whereby the common-mode voltage at the output is regulated to $V_{DD}/2$.

With the resistors $R_{hp,fb} = 10\,k\Omega$ the amplifier A_3 has a bandwidth of $>250\,MHz$ (worst case across technology corner and temperature from $-40\,°C$ to $150\,°C$) and buffers the voltage $V_{out,hp}$ of the high-pass with a gain of $1.2\,dB \approx 1$ and has small negative phase shift of $< 30°$ at $100\,MHz$. Thus, the required demands on the signal compensation of the closed-loop sensing of the Rogowski coil are achieved.

4.2.4   DC Signal Compensation

Each amplifier has an offset voltage due to mismatch of the transistors, which can be described as an effective input offset voltage V_{OS}. Even for a low offset voltage in the range of approximately $1\,mV$, the implemented DC amplification of the forward path $G(s)$ of $76\,dB$ would result in an output voltage $V_{out,R} > 6\,V$ at A_2. Thus, due to the offset voltages of the amplifiers, the output of A_2 would saturate to the supply rails. In consequence, it is essential to compensate for any input offset voltages V_{OS} of the amplifiers. For this purpose, the system is supplemented by a second feedback with a low-pass filter, which is connected in parallel to the high-pass feedback, as shown in Fig. 4.27. The voltage $V_{out,lp}$ at the output of the low-pass filter is buffered by the amplifier A_4, which has an effective voltage gain of $0\,dB$ for the series resistor $R_{coil,s} = 3\,k\Omega$ of the Rogowski coil and feedback resistors $R_{lp,fb} = 100\,k\Omega$. A_4 is implemented, as a symmetrical amplifier according to Appendix, like the amplifiers A_1 and A_2 of the forward path.

The cut-off frequency f_{lp} of the low-pass filter must be in the low mHz region. This results from the lower bandwidth limitation $f_{Rog,min}$ for current sensing, the DC gain $A_{o,G(s)}$ of the forward path $G(s)$ (see Eq. 4.5) and a two-decade safety margin, as shown in Fig. 4.28. This ensures that frequency components $> f_{Rog,min}$ are not attenuated by the low-pass feedback. Therefore, a passive off-chip low-pass filter consisting of $R_{lp} = 100\,k\Omega$ and $C_{lp} = 60\,\mu F$ is proposed.

The transfer behavior of the Rogowski coil, the closed-loop sensor front-end and the resulting overall transfer behavior $V_{out,R}(s)/I_{meas}(s)$ is shown in Fig. 4.28. The low-pass feedback ensures that DC signals are transmitted with approximately $0\,dB$ and thus the offset voltages of the amplifiers are compensated. The forward path $G(s)$ and the high-pass feedback $H(s)$ cause a constant magnitude of the overall transfer behavior in the frequency range $f_{Rog,min}$ to $f_{Rog,max}$. The upper bandwidth limitation $f_{Rog,max}$ is determined by the cut-off frequency f_{hp} of the high-pass filter and the lower bandwidth limitation $f_{Rog,min}$ by $G(s)H(s) = 0\,dB$.

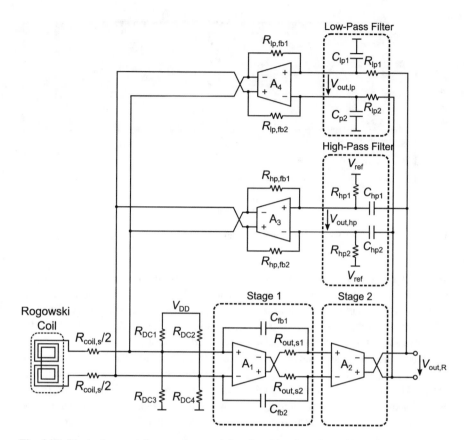

Fig. 4.27 Block diagram of sensor front-end for closed-loop sensing with additional low-pass feedback for offset compensation

4.2.5   Noise Cancelation

For the resolution of the closed-loop sensing, as well as for the open-loop sensing, noise determines the minimum signal current I_{meas}. Therefore, the noise voltage $V_{out,R,noise}$ of the sensor front-end must be as low as possible. As described for open-loop sensing in Sect. 4.1, auto-zeroing or chopping can be used to reduce the dominant $1/f$ noise. Thereby, the noise at low frequencies is ideally reduced down to the thermal noise level of the corresponding circuits. For the closed-loop sensor front-end, both techniques are used for the different transmission paths.

The amplifier A_1 in the forward path $G(s)$ and the amplifier A_3 in the high-pass feedback $H(s)$ amplify signals with frequencies up to 100 MHz. In contrast to open-loop sensing, chopping cannot be used to reduce noise, since here the chopper frequency f_{ch} cannot be selected lower than the signal bandwidth BW_{Rog}. Instead, auto-zeroing is used for these two amplifiers. In order to ensure time-continuous

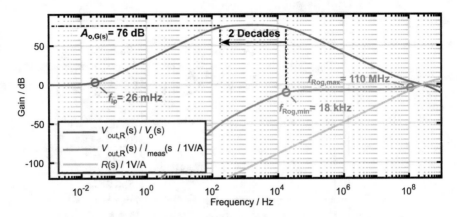

Fig. 4.28 Transfer behavior of the closed-loop sensor front-end

signal amplification, the auto-zeroing ping-pong technique is implemented, as shown in Fig. 4.29. Since the corner frequency of the $1/f$ noise and the flat thermal noise is in the MHz region, a high auto-zeroing frequency f_{az} is required. Because of the limited driving strength of the used amplifiers, f_{az} is limited to $< 5\,\mathrm{MHz}$ for the forward path $G(s)$ and $< 10\,\mathrm{MHz}$ for the feedback path $H(s)$.

Due to the low-pass filter in the low-pass feedback, amplifier A_4 amplifies low-frequency signals up to the kHz range. Thus, noise of A_4 can be area-efficiently reduced by conventional chopping with a chopper frequency in the low MHz range. For this, the chopper frequency $f_{ch} = 3\,\mathrm{MHz}$ is selected in order to reduce the dominant $1/f$ noise. The required chopper low-pass filter for ripple reduction and noise filtering must be designed to have no influence on the feedback signal. To ensure that signals up to $f_{Rog,min} = 18\,\mathrm{kHz}$ are not attenuated by the chopper low-pass filter, its cut-off frequency $f_{lp,ch} = 237\,\mathrm{kHz}$ is selected with a safety margin of one decade. This is achieved with the filter consisting of $R_{lp,ch} = 12.8\,\mathrm{k\Omega}$ and $C_{lp,ch} = 52.4\,\mathrm{pF}$.

The capacitance $C_{lp,ch}$ of the chopper low-pass filter and the series resistance $R_{coil,s}$ of the Rogowski coil form a parasitic low-pass filter with a cut-off frequency of approximately $1\,\mathrm{MHz}$. To ensure that this filter has no effect on bandwidth of current sensing, the resistors $R_{lp,s} = 50\,\mathrm{k\Omega}$ are connected in series to the output of the chopper low-pass filter. As a result, the low-pass feedback is attenuated, but this can be compensated by a higher gain of the amplifier A_4.

By combining these noise reduction techniques, the output noise of the sensor front-end can be reduced by 97 % and an effective noise voltage $V_{out,R,noise}$ of 20.4 mVrms can be achieved.

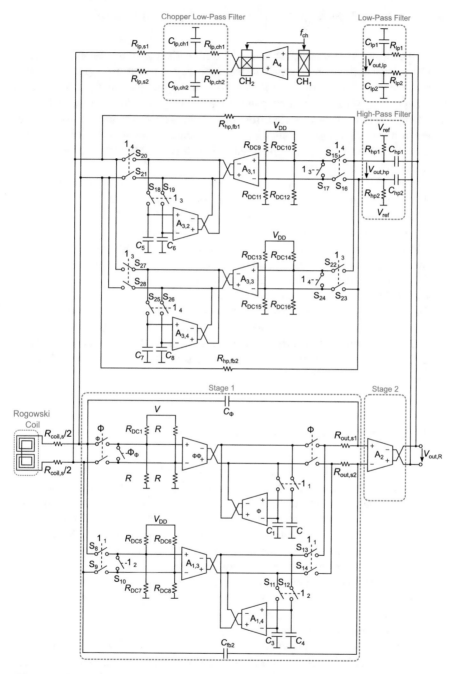

Fig. 4.29 Block diagram of sensor front-end for closed-loop sensing including noise reduction techniques

4.2.6   Overall System

With the closed-loop sensor front-end, the differentiating transmission behavior of
the Rogowski coil is compensated. In the frequency range from 17 kHz to 63 MHz
a high sensitivity $S_{out,R}$ of -7.0 dB related to the signal current I_{meas} for on-chip
sensing is achieved, which corresponds to 447 mV/A, as shown in Fig. 4.30. The
bandwidth is lower than expected from the previous considerations in this section,
which is due to the implementation of noise reduction techniques. Nevertheless, the
achieved bandwidth exceeds the required bandwidth (>50 MHz) derived for on-chip
current sensing in Sect. 2.1.1.

Figure 4.31 shows the transient signals of the closed-loop on-chip current sensing
with a planar Rogowski coil for a 5 MHz sinusoidal signal current I_{meas} with a peak-
to-peak current of 0.92 A. The sensing output $V_{out,R}$ has a peak-to-peak voltage of
413 mV, which corresponds to a sensitivity $S_{out,R}$ of 447 mV/A.

Figure 4.32 shows the response for the closed-loop on-chip current sensing for
a saw tooth signal current I_{meas} with an amplitude of 1 A at a switching frequency
of 2.4 MHz. The fast current changes, due to the low duty cycle $D = 0.125$, can be
reproduced due to the high total bandwidth.

In conclusion, with the presented closed-loop sensor front-end, the required
bandwidth for on-chip current sensing (>50 MHz) and off-chip current sensing
(>10 MHz) is achieved, as derived in Sects. 2.1.1 and 2.1.2. Due to the combination
of DC and AC signal compensation a high sensitivity of 447 mV/A with a low
output offset of <1.5 mV can be achieved.

4.3   Open-Loop vs. Closed-Loop Sensing

In the following, the implemented sensor front-ends for open-loop current sensing in
Sect. 4.1 and closed-loop current sensing in Sect. 4.2 are compared with each other.

Table 4.4 shows an overview of different parameters of the sensor front-ends. The
sensitivity of the closed-loop sensor front-end with 447 mV/A is 10.2 times higher
than the sensitivity of the open-loop sensor front-end with 43.7 mV/A. However,
due to the much more complex circuitry and the use of multiple amplifiers for AC
and DC signal compensation, the effective noise voltage $V_{out,R,noise}$ increases at
the same time. Since the noise increases more than the sensitivity, the closed-loop
sensor front-end achieves a lower SNR compared to the open-loop sensor front-end
(-8 dB). This increases the minimum detectable current for closed-loop sensing to
132 mArms (with the open-loop sensor front-end 50 mArms can be measured). Both
sensor front-ends reach the required bandwidth >50 MHz for fast on-chip current
sensing as derived in Sect. 2.1.1. Due to the DC signal compensation of the closed-
loop sensor front-end, the resulting offset voltage $V_{out,R,OS}$ at the output can be
reduced by a factor of 69 compared to the open-loop sensor front-end. Closed-loop

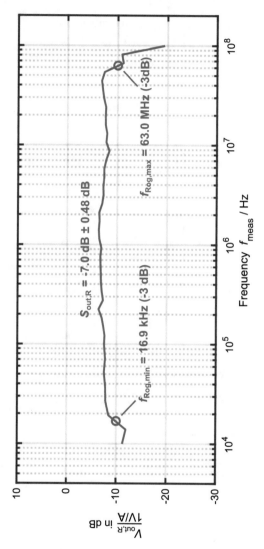

Fig. 4.30 Transfer behavior of closed-loop on-chip current sensing with a planar Rogowski coil

Fig. 4.31 Transient
measurement result of
closed-loop on-chip current
sensing at f_{meas}=5 MHz

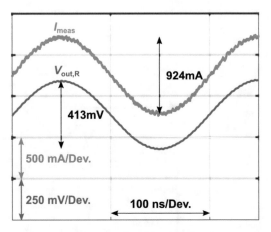

Fig. 4.32 Transient response
of closed-loop on-chip
current sensing with a 1 A
saw tooth signal current at a
switching frequency of
2.4 MHz

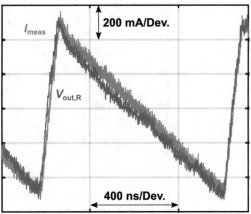

Table 4.4 Parameter comparison between open-loop and closed-loop sensing

Parameter	Symbol	Open-loop	Closed-loop
Sensitivity	S_{Rog}	43.7 mV/A	447 mV/A
Noise	$V_{out,R,noise}$	2.2 mV	59.6 mV
Signal to noise ratio	SNR	⟵ −8 dB	
Minimal signal current	$I_{meas,min}$	50 mArms	132 mArms
Bandwidth	BW	15.1 kHz–74.7 MHz	16.9 kHz–63 MHz
Output offset	$V_{out,R,OS}$	104 mV	<1.5 mV
Area	A	0.89 mm^2	1.2 mm^2

sensing is more complex but does not require a significantly larger area compared to
the open-loop sensor front-end. This is because the very large integration capacitors
C_{int1} (Fig. 4.3) can be omitted.

Due to the differentiating transfer behavior of the Rogowski coil, large amplitudes are generated at the output of the Rogowski coil at high signal frequencies.

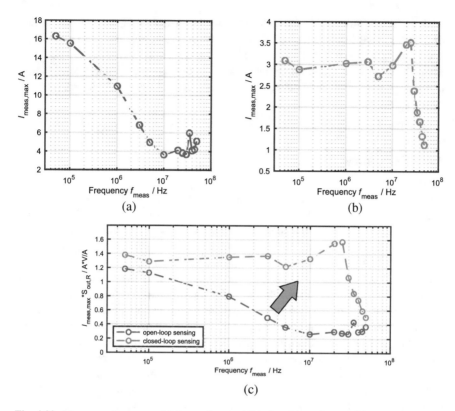

Fig. 4.33 Linear sensing range of (**a**) open-loop and (**b**) closed-loop front-end for on-chip current sensing depending on signal frequency. A comparison considering the sensitivity of the different front-ends is shown in (**c**)

As a result, the sensor front-end for open-loop sensing reaches the non-linear amplification range (defined by the 1 dB compression point, see Fig. 4.18) earlier at high signal frequencies f_{meas}. The maximum linear amplitude sensing range $I_{meas,max}$ decreases with f_{meas}, as shown in Fig. 4.33a. In contrast, the closed-loop sensor front-end compensates increasing amplitudes at higher frequencies and a constant sensing range $I_{meas,max}$ is achieved over a wide frequency range, as shown in Fig. 4.33b. For a fair comparison of these two sensor front-ends in Fig. 4.33c, the respective linear sensing range $I_{meas,max}$ is multiplied by the corresponding sensitivity $S_{out,R} = V_{out,R}/I_{meas}$ from Table 4.4. Because of the decreasing loop gain of the closed-loop front-end at high frequencies, signal currents >25 MHz cannot be attenuated sufficiently, which reduces also the sensing range of the closed-loop front-end. However, the closed-loop sensor front-end has a more constant sensing range, which is a significant improvement over the open-loop sensor front-end.

As a further advantage of the closed-loop sensor front-end, the complex tuning of the pole and the zero in the open-loop sensor front-end (see Sect. 4.1) is no longer necessary, since the integrating transfer behavior is realized with a passive high-pass filter. With the closed-loop implementation, only the variation of the absolute values of the high-pass filter must be adjusted to get a cut-off frequency f_{hp} of 100 MHz (see Sect. 4.2).

4.4 Summary

The Rogowski coil has a differentiating transfer behavior. Therefore, the output voltage is proportional to the dI/dt and has a phase shift of $+90°$ to the signal current to be measured. Both dependencies can be compensated with an integrating transfer characteristic of the sensor front-end to reflect the waveform of the signal current. In this work, an open-loop sensor front-end is investigated for on-chip current sensing and off-chip current sensing. The combination of two integrator stages in the open-loop front-end allows chopping for low-frequency errors cancelation. As a major advantage of the two-stage integrator circuit chopping can be applied for offset and noise reduction and, in contrast to conventional chopping, the chopper frequency can be lower than the signal bandwidth. Experimental results confirm that chopping related frequencies >5 MHz disappears in the noise floor and an output noise voltage of 2.2 mArms is reached. Experimental results confirm that frequency components 15 kHz $< f_{meas} <$ 75 MHz can be measured with a sensitivity of 43.7 mV/A by the implemented planar Rogowski coil [8] and frequency components 16 kHz $< f_{meas} <$ 15 MHz can be measured with a sensitivity of 3.1 mV/A by the implemented helix-shaped Rogowski coil [9]. The bandwidth of 75 MHz for on-chip current sensing with a planar Rogowski coil exceeds the state-of-the-art current sensing with a fully integrated coil by a factor of 25. The bandwidth of 15 MHz for off-chip current sensing with a helix-shaped Rogowski coil exceeds the state of the art by a factor of 5. The implemented Rogowski coils with the two-stage integrator circuit cover the required bandwidth for on-chip and off-chip current sensing and is suitable for real-time current sensing. This was confirmed by transient measurements with sinusoidal signal current up to 70 MHz and current pulses of 60 A and a regulated slope of 1 kA/μs.

Based on the implemented open-loop sensing concept, this book introduces the first known compensated sensor front-end for a Rogowski coil. Experimental results confirm that the closed-loop sensor front-end achieves a significantly higher sensitivity of 447 mV/A, a larger linear sensing range and a smaller output offset of <1.5 mV with a comparable bandwidth compared to the open-loop front-end. The output signal of the Rogowski coil is damped by the feedback in the form of a high-pass filter. The damping of the Rogowski coil at high frequencies is achieved by a pseudo-differential amplifier with a bandwidth of >250 MHz. This allows to extend the linear sensing range significantly at high frequencies. Furthermore, the output offset is improved by an additional low-pass feedback by a factor of 69 and reaches <1.5 mV compared to open-loop current sensing.

Table 4.5 shows a comparison of the introduced on-chip current sensing with the state-of-the-art sensor types. The sense-FET implementation in [19] reaches a bandwidth of 60 MHz, but is only suitable for low-voltage applications (<4 V). A sense-FET current sensing circuit for high-voltage switching applications up to 40 V is published in [20], but has a limited bandwidth of 7 MHz. [21] covers a large input range of the current to be measured, but has a bandwidth in the kHz region. A digitization of the current signal is done in [22], but only reaches a bandwidth of 0.2 kHz. The introduced current sensors in this work achieve the highest bandwidth with a comparable input range and chip area. The closed-loop implementation reaches a high sensitivity of 447 mV/A and furthermore provides a constant input range over a wide frequency range.

A comparison of the introduced off-chip current sensing with the state of the art is shown in Table 4.6. The chip with a total thickness of 60 μm is mounted directly on a PCB for contactless sensing of the current through the underlying power line and achieves the highest bandwidth of 15 MHz, which exceeds the state of the art by factor 5. [23] achieves a highly sensitive current sensing and an on-chip digitalization of the signal current, but has a bandwidth of 0.1 kHz and no galvanic isolation. A required galvanic isolation and a large input range for power applications are provided in [7, 24, 25], but the bandwidth is not sufficient for fast switching applications.

Appendix

Fully Differential Symmetrical Amplifier

The implementation of the fully differential symmetrical amplifier for signal processing is shown in Fig. 4.34. For high bandwidth optimization, the current mirrors M_5 to M_7 and M_4 to M_6 are implemented with a ratio $n = 5$. Furthermore, the output stage is optimized for low parasitic capacitances. Therefore, the source-follower M_{F1} and M_{F2} buffers the output of A_2 for the required CMFB. This enables a nominal DC gain of 40 dB with a bandwidth of 26 MHz.

Stability Analysis of Rogowski Coil Closed-Loop Sensing

Due to the increased resistance of the Rogowski coil in the closed-loop current sensing, the system becomes unstable. By adding a series resistor $R_{out,s}$ to the output of the amplifier A_1 of Stage 1, the stability of the system can be ensured. Figure 4.35

Table 4.5 Comparison of CMOS on-chip current sensing with prior art

	[21]	[22]	[19][d]	[20][d]	This work open-loop	This work closed-loop
Sensor type	Shunt	Shunt	Sense-FET	Sense-FET	Rogowski coil	Rogowski coil
Process node	0.13 μm	0.13 μm	0.5 μm	0.18 μm	0.18 μm	0.18 μm
Area	0.39 mm^2	1.1 mm^2	n.r.[a]	n.r.[a]	1.24 mm^2	1.55 mm^2
BW	DC–25 kHz	DC–0.2 kHz	DC–60 MHz	DC–7 MHz	15 kHz–75 MHz	16.9 kHz–63 MHz
Input range	0.1 A–100 A	1 A	5 A	60 mA–1.5 A	±4.1 A (±40 A[c])	±3.5 A
P_{diss}	15.6 mW	8 μW[b]	n.r.[a]	n.r.[a]	2.7 mW	27.9 mW
Output	Analog	Digital	Analog	Analog	Analog	Analog
Sensitivity	n.r.[a]	50 mA	n.r.[a]	n.r.[a]	43.7 mV/A	447 mV/A
$I_{meas,min}$	n.r.[a]	n.r.[a]	n.r.[a]	n.r.[a]	0.05 Arms	0.08 Arms

[a]Not reported
[b]Calculated
[c]Extrapolated
[d]Simulation only

Table 4.6 Comparison of CMOS off-chip current sensing with prior art

	[23]	[24]	[25]	[7]	This work open-loop
Sensor type	Shunt	Hall	Fluxgate	Hall + coil	Rogowski coil
Process node	0.13 μm	0.18 μm	0.35 μm	0.18 μm	0.18 μm
Area	0.85 mm²	8.75 mm²	n.r.[a]	8.75 mm²	1.64 mm²
BW	DC−0.1 kHz	DC−400 kHz	DC−47 kHz	DC−3 MHz	15 kHz−15 MHz
Input range	±5 A	n.r.[a]	100 μA−80 A	±18 A	±60 A
P_{diss}	0.02 mW	40 mW[b]	23/17 mW	38.5 mW[b]	2.7 mW
Output	Digital	Analog	Digital	Analog	Analog
Sensitivity	200 μA	80 mV/A	n.r.[a]	n.r.[a]	3.1 mV/A
$I_{meas,min}$	n.r.[a]	n.r.[a]	n.r.[a]	0.48 Arms	0.71 Arms

[a]Not reported
[b]Calculated

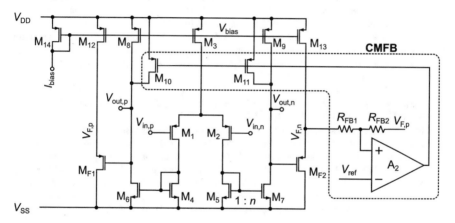

Fig. 4.34 Schematic of a fully differential symmetrical amplifier

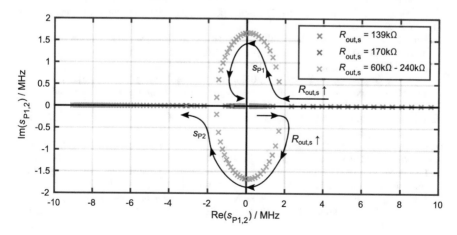

Fig. 4.35 Simulated poles of $V_{out,S1}(s)/V_{coil}(s)$ on transistor level for different values of $R_{out,s}$

shows the simulated poles $s_{p1,2}$ on transistor level for different values of $R_{out,s}$. It can be seen that for $R_{out,s} \geq 139\,\Omega$, the poles have a negative real part and for $R_{out,s} \geq 170\,\Omega$ additionally lie on the real axis. Therefore, $R_{out,s} = 170\,\Omega$ is chosen for the implementation in the closed-loop sensor front-end for the Rogowski coil.

References

1. Han, R.-Y., Wu, J.-W., Ding, W.-D., Jing, Y., Zhou, H.-B., Liu, Q.-J., et al. (2015). Hybrid PCB Rogowski coil for measurement of nanosecond-risetime pulsed current. *IEEE Transactions on Plasma Science, 43*(10), 3555–3561. ISSN: 0093-3813. https://doi.org/10.1109/TPS.2015.2415517.
2. Limcharoen, W., & Yutthagowith, P. (2013). Rogowski coil with an active integrator for measurement of switching impulse current. In *Proceedings of the Telecommunications and Information Technology 2013 10th International Conference on Electrical Engineering/Electronics, Computer* (pp. 1–4). https://doi.org/10.1109/ECTICon.2013.6559578.
3. Yutthagowith, P., & Leelachariyakul, B. (2014). A Rogowski coil with an active integrator for measurement of long duration impulse currents. In *Proceedings of the International Conference on Lightning Protection (ICLP)* (pp. 1761–1765). https://doi.org/10.1109/ICLP.2014.6973414.
4. Liu, Y., Lin, F., Zhang, Q., & Zhong, H. (2011). Design and construction of a Rogowski coil for measuring wide pulsed current. *IEEE Sensors Journal, 11*(1), 123–130. ISSN: 1530-437X. https://doi.org/10.1109/JSEN.2010.2052034.
5. Ray, W. F., & Hewson, C. R. (2000). High performance Rogowski current transducers. In *Proceedings of the Conference Record of the 2000 IEEE Industry Applications Conference. Thirty-Fifth IAS Annual Meeting and World Conf. Industrial Applications of Electrical Energy (Cat. No.00CH37129)* (Vol. 5, pp. 3083–3090). https://doi.org/10.1109/IAS.2000.882606.
6. Wang, B., Wang, D., & Wu, W. (2009). A Rogowski coil current transducer designed for wide bandwidth current pulse measurement. In *Proceedings of the IEEE 6th International Power Electronics and Motion Control Conference* (pp. 1246–1249). https://doi.org/10.1109/IPEMC.2009.5157575.
7. Jiang, J., & Makinwa, K. (2016). A hybrid multipath CMOS magnetic sensor with $210\,\mu$Trms resolution and 3 MHz bandwidth for contactless current censing. In *Proceedings of the IEEE International Solid-State Circuits Conference (ISSCC)* (pp. 204–205). https://doi.org/10.1109/ISSCC.2016.7417978.
8. Funk, T., & Wicht, B. (2018). A fully integrated DC to 75 MHz current sensing circuit with on-chip Rogowski coil. In *Proceedings of the IEEE Custom Integrated Circuits Conference (CICC)* (pp. 1–4). https://doi.org/10.1109/CICC.2018.8357028.
9. Funk, T., Groeger, J., & Wicht, B. (2019). An integrated and galvanically isolated DC-to-15.3 MHz hybrid current sensor. In *Proceedings of the IEEE Applied Power Electronics Conf. and Exposition (APEC)* (pp. 1010–1013).
10. Witte, F., Makinwa, K., & Huijsing, J. (2009). *Dynamic offset compensated CMOS amplifiers*. Berlin: Springer.
11. Oliver, B. M. (1965). Thermal and quantum noise. *Proceedings of the IEEE, 53*(5), 436–454. ISSN: 0018-9219. https://doi.org/10.1109/PROC.1965.3814.
12. Leach, W. M. (1994). Fundamentals of low-noise analog circuit design. *Proceedings of the IEEE, 82*(10), 1515–1538. ISSN: 0018-9219. https://doi.org/10.1109/5.326411.
13. Rong, W., Huijsing, J. H., & Makinwa, K. A. (2013). *Precision instrumentation amplifiers and read-out integrated circuits*. Berlin: Springer. ISBN: 9781461437307. https://doi.org/10.1007/978-1-4614-3731-4.

14. Enz, C. C., & Temes, G. C. (1996). Circuit techniques for reducing the effects of op-amp imperfections: Autozeroing, correlated double sampling, and chopper stabilization. *Proceedings of the IEEE, 84*(11), 1584–1614. ISSN: 0018-9219. https://doi.org/10.1109/5.542410.

15. Pertijs, M. A. P., & Kindt, W. J. (2010). A 140 dB-CMRR current-feedback instrumentation amplifier employing ping-pong auto-zeroing and chopping. *IEEE Journal of Solid-State Circuits, 45*(10), 2044–2056. ISSN: 0018-9200. https://doi.org/10.1109/JSSC.2010.2060253.

16. Tang, A. T. K. (2002). A 3 μV-offset operational amplifier with 20 nV/$\sqrt{Hz}$ input noise PSD at DC employing both chopping and autozeroing. In *Proceedings of the IEEE International Solid-State Circuits Conference. Digest of Technical Papers (Cat. No.02CH37315)* (Vol. 1, pp. 386–387). https://doi.org/10.1109/ISSCC.2002.993094.

17. Xue, Y., Lu, J., Wang, Z., Tolbert, L. M., Blalock, B. J., & Wang, F. (2014). A compact planar Rogowski coil current sensor for active current balancing of parallel-connected silicon carbide MOSFETs. In *Proceedings of the IEEE Energy Conversion Congress and Exposition (ECCE)* (pp. 4685–4690). https://doi.org/10.1109/ECCE.2014.6954042.

18. Mohieldin, A. N., Sanchez-Sinencio, E., & Silva-Martinez, J. (2003). A fully balanced pseudo-differential OTA with common-mode feedforward and inherent common-mode feedback detector. *IEEE Journal of Solid-State Circuits, 38*(4), 663–668. ISSN: 0018-9200. https://doi.org/10.1109/JSSC.2003.809520.

19. Jiang, Y., Swilam, M., Asar, S., & Fayed, A. (2018). An accurate sense-FET-based inductor current sensor with wide sensing range for buck converters. In *Proceedings of the IEEE International Symposium on Circuits and Systems (ISCAS)* (pp. 1–4). https://doi.org/10.1109/ISCAS.2018.8351083.

20. Renz, P., Lamprecht, P., Teufel, D., & Wicht, B. (2016). A 40 V current sensing circuit with fast on/off transition for high-voltage power management. In *Proceedings of the IEEE 59th International Midwest Symposium on Circuits and Systems (MWSCAS)* (pp. 1–4). https://doi.org/10.1109/MWSCAS.2016.7870011.

21. Rothan, F., Lhermet, H., Zongo, B., Condemine, C., Sibuet, H., Mas, P., et al. (2011). A ±1.5% nonlinearity 0.1-to-100 A shunt current sensor based on a 6 kV isolated micro-transformer for electrical vehicles and home automation. In *Proceedings of the IEEE International Solid-State Circuits Conference* (pp. 112–114). https://doi.org/10.1109/ISSCC.2011.5746242.

22. Shalmany, S. H., Draxelmayr, D., & Makinwa, K. A. A. (2013). A micropower battery current sensor with ±0.03%(3σ) inaccuracy from −40 to +85°C. In *Proceedings of the IEEE International Solid-State Circuits Conference on Digest of Technical Papers* (pp. 386–387). https://doi.org/10.1109/ISSCC.2013.6487781.

23. Shalmany, S. Heidary, Draxelmayr, D., & Makinwa, K. A. A. (2017). A ±36-A integrated current-sensing system with a 0.3% gain error and a 400 μA offset from −55°C to +85°C. *IEEE Journal of Solid-State Circuits, 52*(4), 1034–1043. ISSN: 0018-9200. https://doi.org/10.1109/JSSC.2016.2639535.

24. Jiang, J., & Makinwa, K. A. A. (2015). A multi-path CMOS hall sensor with integrated ripple reduction loops. In *Proceedings of the IEEE Asian Solid-State Circuits Conference (A-SSCC)* (pp. 1–4). https://doi.org/10.1109/ASSCC.2015.7387504.

25. Snoeij, M. F., Schaffer, V., Udayashankar, S., & Ivanov, M. V. (2016). Integrated fluxgate magnetometer for use in isolated current sensing. *IEEE Journal of Solid-State Circuits, 51*(7), 1684–1694. ISSN: 0018-9200. https://doi.org/10.1109/JSSC.2016.2554147.

Chapter 5
Hall Current Sensor

A Hall sensor is well suited to extend the limited low-frequency range of a Rogowski coil towards DC signal currents. It provides an adjustable sensitivity to adapt it to the Rogowski path and it also offers a galvanic isolation for power electronics applications.

In this book, different Hall devices for on-chip and off-chip current sensing are proposed. Section 5.1 explains the operation principle of a Hall device and the resulting design dependencies in terms of sensitivity. The design of a planar Hall device for on-chip current sensing is presented in Sect. 5.2 [1], while Sect. 5.3 covers the design of a vertical Hall device for off-chip current sensing [2].

5.1 Operation Principle

The Hall effect enables sensing of the magnetic field $\vec{B}$ of the signal current to be measured in the form of a voltage. When a magnetic field penetrates a current-carrying conductor, the charge q, which moves with the drift velocity $\vec{v}$ in the conductor, is deflected by the Lorenz force

$$\vec{F_L} = q \cdot \vec{v} \times \vec{B}. \tag{5.1}$$

This results in the potential difference V_{Hall}, which is a measure for the magnetic field $\vec{B}$.

Figure 5.1 shows a simplified Hall device with a magnetic field $\vec{B}$ in positive z-direction. A bias current $I_{bias,H}$ is fed in between terminals A and C and the Hall voltage V_{Hall} is measured between terminal B and D. For a negative charge $q_n = -q$ (i.e., a single electron) with $\mathbf{v} = (0, -v_y, 0)$ and $\vec{B} = (0, 0, B_z)$, the Lorenz force has only components in x-direction and Eq. 5.1 can be simplified to

© The Editor(s) (if applicable) and The Author(s), under exclusive license to
Springer Nature Switzerland AG 2020
T. Funk, B. Wicht, *Integrated Wide-Bandwidth Current Sensing*,
https://doi.org/10.1007/978-3-030-53250-5_5

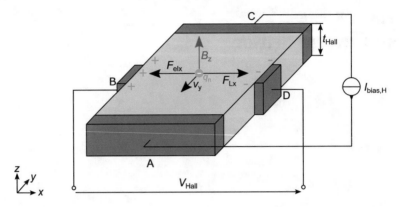

Fig. 5.1 Functionality of a Hall device

$$F_{Lx} = q_n \cdot v_y \cdot B_z. \tag{5.2}$$

Due to the deflection of the negative charge q_n a positive potential difference between terminals B and D is created and an electric field is generated. Because of this electric field, an electric force F_{elx} acts on q_n with the strength $F_{elx} = -F_{Lx}$. With consideration of the geometry and the dopant density N_{dop} of an integrated Hall device, the Hall voltage V_{Hall} can be calculated by

$$V_{Hall} = -\frac{1}{N_{dop} \cdot q} \cdot \frac{I_{bias,H} \cdot B_z}{t_{Hall}}, \tag{5.3}$$

where t_{Hall} is the thickness of the Hall device, respectively, the depth of the active N-well in the used semiconductor. The first term is referred to as Hall coefficient R_H. For semiconductors, R_H consists of the reciprocal of the doping density N_{dop} and the charge q. Furthermore, it is extended by the Hall factor r_H according to Popovic [3]:

$$R_H = -\frac{r_H}{N_{dop} \cdot q}. \tag{5.4}$$

The Hall factor r_H depends on doping and temperature and also takes into account the interaction of the charge carriers with the crystal lattice. Typically r_H is in the range of $1.0 \ldots 1.2$ at $300\,K$ (applies to silicon semiconductors) [4–6]. According to [3] this results in the Hall voltage

$$V_{Hall} = R_H \cdot \frac{I_{bias,H} \cdot B_z}{t_{Hall}} = -\frac{r_H}{N_{dop} \cdot q} \cdot \frac{I_{bias,H} \cdot B_z}{t_{Hall}}. \tag{5.5}$$

For current sensing, limitations such as bandwidth or noise can be neglected. The bandwidth depends on the parasitic junction capacitance C_{np} between the active N-well and the substrate and is typically in the order of several hundred MHz. The input capacitance of the subsequent sensor front-end reduces the noise of the Hall device to a level of a few μV.

As a conclusion, a Hall device can measure the strength of a magnetic field $\vec{B}$ perpendicular to the orientation of bias current $I_{bias,H}$ and the terminals for measuring the Hall voltage V_{Hall}. V_{Hall} is proportional to the magnetic field $\vec{B}$ and $I_{bias,H}$. V_{Hall} also depends on parameters of the used CMOS technology. To achieve a large Hall voltage V_{Hall} according Eq. 5.5, the sensor requires a low dopant density N_{dop} and a small thickness t_{Hall} of the active N-well. For typical technology values, as described later in Table 5.1 in Sect. 5.2, a bias current $I_{bias,H}$ of 1 mA and a magnetic field $\vec{B}$ of 1 mT, Eq. 5.5 results in a Hall voltage V_{Hall} of 69 μV.

5.2 Hall Device for On-Chip Current Sensing

The need for a fully integrated sensing of DC currents was derived in Sect. 2.1.1. To use the properties of a Hall sensor for on-chip current sensing most effectively, a cross-shaped Hall device is placed besides an on-chip power line to detect the magnetic field $\vec{B}$ in the z-direction [1]. By optimizing the geometry of the cross-shaped Hall device in Fig. 5.2, a sensitivity of 24 μV/A is achieved.

In addition to the dopant density N_{dop} and thickness t_{Hall} of the active N-well of the Hall device in Eq. 5.5, the Hall voltage is also influenced by the geometric shape of the device. Therefore, the correction factor G is introduced to take into account the efficiency of the specific shape of the Hall device [3, 4]. The Hall voltage V_{Hall} of Eq. 5.5 is expanded by G and can be calculated by

$$V_{Hall} = G \cdot R_H \cdot \frac{I_{bias,H} \cdot B_z}{t_{Hall}}. \tag{5.6}$$

The ideal maximum value of the geometry factor is $G_{ideal} = 1$. For different shapes and dimensions G varies between 0 and 1. A highly symmetrical shape and uniform thickness of the plate are usually chosen, since this symmetry can be used to cancel the offset voltage of the sensor itself, as will be investigated in Sect. 6.1. Typical symmetrical geometries for Hall devices are shown in Fig. 5.2.

The calculation of the geometry factor G according [3] of these geometries is described in Appendix. Thereby G only depends on the size of the W_{Hall} and L_{Hall}. Figure 5.3 shows the resulting value of G for different ratios L_{Hall}/W_{Hall}. The cross-shaped geometry reaches the best value ($G > 0.99$) over a wide range of $L_{Hall}/W_{Hall} = 0 \ldots 10$.

The geometry factor G for cross-shaped Hall devices and different ratios L_{Hall}/W_{Hall} is verified by Silvaco TCAD simulations. Figure 5.4 shows the simplified cross-shaped layout with the terminals A to D and the required dimensions for the TCAD simulation. The Hall device is simulated at semiconductor level with a defined magnetic field $\vec{B}$ in z-direction and bias currents $I_{bias,H}$. The sensitivity

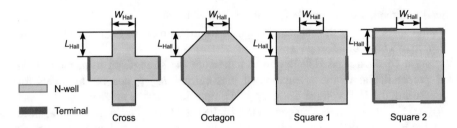

Fig. 5.2 Different shapes of integrated Hall devices [3]

Fig. 5.3 Calculated geometry factor G for different shapes of Hall devices

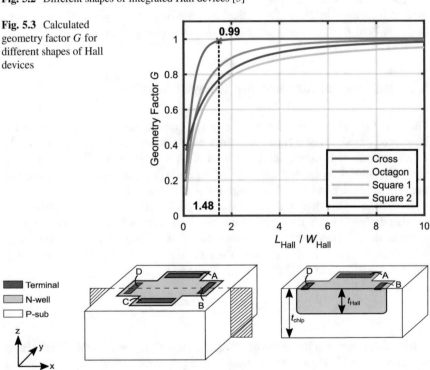

Fig. 5.4 Cross-section of an integrated cross-shaped Hall device

$$S_A = \frac{V_{Hall}}{B_z} = G \cdot R_H \cdot \frac{I_{bias,H}}{t_{Hall}} \qquad (5.7)$$

and therefore the influence of G can be extracted from the simulated Hall voltage V_{Hall}.

The parameters for these simulations are shown in Table 5.1. Typically values are chosen for the dopant density of the P-substrate and the active N-well [7]. The exact value of the thickness t_{Hall} of the N-well was provided by the vendor of the used technology. The thickness of the chip was set to $t_{chip} = 5\,\mu m$, which is very

Table 5.1 Parameter for TCAD simulation

Parameter	Symbol	Value
Dopant density P-substrate (Boron)	P_{dop}	$1 \cdot 10^{15} \text{cm}^{-3}$
Dopant density N-well (Arsenic)	N_{dop}	$1 \cdot 10^{17} \text{cm}^{-3}$
Thickness P-substrate (Boron)	t_{chip}	$5\,\mu m$
Thickness N-well (Arsenic)	t_{Hall}	$1\,\mu m$
Hall factor	r_H	1.1

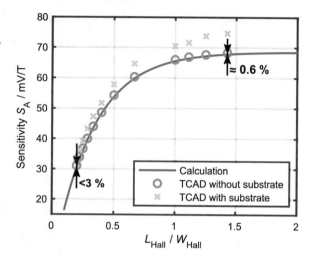

Fig. 5.5 Comparison between the calculated and simulated absolute sensitivity S_A for different dimensions of the cross-shaped Hall device for $I_{bias,H} = 1$ mA

small compared to the normal thickness of 250 μm, but leads to the same sensitivity S_A at a much faster simulation time. Figure 5.5 shows the calculated and simulated absolute sensitivity S_A of the cross-shaped Hall device for different dimensions. As expected from the increasing geometry factor G in Fig. 5.3, the sensitivity increases for an increasing L_{Hall}/W_{Hall} ratio. For verification of the calculated values, the Hall device was also simulated without considering the substrate. The calculated and simulated sensitivity matches with an error <3 %. For large L_{Hall}/W_{Hall} the error can be reduced to approximately 0.6 %. By considering the substrate, a higher sensitivity S_A is reached. This results from the depletion region between the substrate and active N-well. Thereby, the effective thickness of the N-well is getting smaller which results in a higher sensitivity S_A, see Eq. 5.7. Additionally, W_{Hall} gets smaller, which leads to a higher geometry factor G as shown in Fig. 5.3. Therefore, a maximum sensitivity S_A of 75 mV/T is achieved for $L_{Hall}/W_{Hall} = 1.4$.

The influence of the depletion region can also be confirmed by the Hall resistance R_{Hall} between the terminals A & C and B & D, respectively, in Fig. 5.6a. Figure 5.6b shows R_{Hall} for different L_{Hall}/W_{Hall} ratios and the influence of the substrate. The increase in R_{Hall} is also a consequence of the smaller effective dimensions of the N-well. For larger L_{Hall}/W_{Hall} the difference by considering the P-substrate increases from 80 Ω to 148 Ω.

Based on the previous considerations, a cross-shaped Hall device was implemented in a 180 nm HV CMOS technology. To achieve the highest simulated

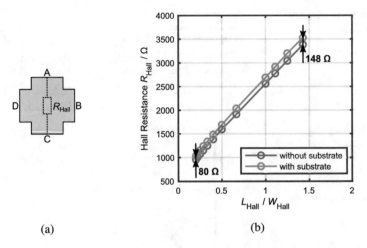

Fig. 5.6 Parasitic Hall resistance R_{Hall} **(a)** in a cross-shaped Hall device and **(b)** TCAD simulation results for different dimensions for $I_{bias,H} = 1\,mA$

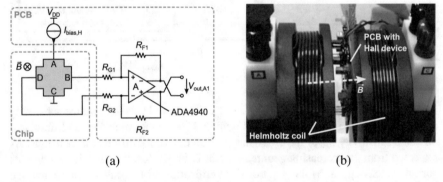

Fig. 5.7 Measurement setup for planar Hall device characterization. **(a)** schematic of read out circuit **(b)** photograph of Helmholtz coil with PCB

absolute sensitivity $S_A = 75\,mV/T$ in Fig. 5.5, the Hall device was implemented area-efficiently with $W_{Hall} = 14\,\mu m$ and $L_{Hall} = 20\,\mu m$.

For the characterization of sensitivity S_A the bias current $I_{bias,H}$ for the Hall device is generated by an external current source and the Hall voltage V_{Hall} is amplified by the discrete low noise, low distortion fully differential amplifier ADA4940 from Analog Devices with a DC gain $A_o = 100$, see Fig. 5.7a. The magnetic field $\vec{B}$ is generated by a Helmholtz coil, which makes it possible to generate homogeneous and constant magnetic fields up to 20 mT [8]. A photograph of the measurement setup including the Helmholtz coil and the PCB with the test chip is shown in Fig. 5.7b.

For verification of the TCAD simulation, the simulated absolute sensitivity S_A and the Hall resistance R_{Hall} between the terminals A & C and B & D, respectively,

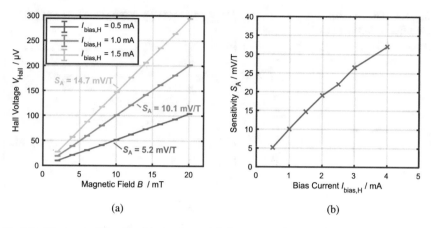

Fig. 5.8 Measurement results of the implemented planar cross-shaped Hall device: **(a)** mean value and standard deviation of Hall voltage V_{Hall} and **(b)** sensitivity S_A for different bias currents $I_{\text{bias,H}}$

Table 5.2 Comparison between TCAD simulation and measurement results of the implemented cross-shaped Hall device

Parameter	Symbol	TCAD simulation	Measurement
Sensitivity @ $I_{\text{bias,H}} = 1\,\text{mA}$	S_A	$75\,\text{mV/T}$	$10.1\,\text{mV/T}$
Resistance	R_{Hall}	$3532\,\Omega$	$1355\,\Omega$

are compared with the measurement results. As expected from Eq. 5.6, the measured Hall voltage V_{Hall} increases linearly with the magnetic field $\vec{B}$ and bias current $I_{\text{bias,H}}$. Figure 5.8 confirms this relationship. Nevertheless, the sensitivity S_A is significantly lower than expected by the TCAD simulation, as shown in Table 5.2. This results from the assumptions of the dopant densities P_{dop} and N_{dop} of the substrate, respectively, the active N-well, and of the Hall factor r_H in Table 5.1. The measured Hall resistance $R_{\text{Hall}} = 1355\,\Omega$ is also significantly lower as expected from the TCAD simulation, see Table 5.2. This is only related to the assumed dopant density N_{dop} of the active N-well.

To align the assumed dopant density N_{dop} in the TCAD simulation to the implemented chip, first the Hall resistance R_{Hall} will be considered. R_{Hall} of a cross-shaped Hall device is approximated by Xu and Pan [9]

$$R_{\text{Hall}} = \frac{1}{\mu_{\text{Nwell}} \cdot q \cdot N_{\text{dop}} \cdot t_{\text{Hall}}} \cdot \left(\frac{2L_{\text{Hall}}}{W_{\text{Hall}}} + \frac{2}{3} \right). \tag{5.8}$$

The carrier mobility μ_{Nwell} is a function of N_{dop}. Therefore, N_{dop} can be changed to adjust the simulated Hall resistance R_{Hall} to the measurements.

Figure 5.9 shows the simulated Hall resistance R_{Hall} of the implemented Hall device for different N_{dop}. With an interpolated dopant density $N_{\text{dop}} = 4.62 \cdot 10^{17}\,\text{cm}^{-3}$ the simulated R_{Hall} corresponds to the measurement results. The absolute

Fig. 5.9 Hall resistance R_{Hall}
for different dopant densities
N_{dop}

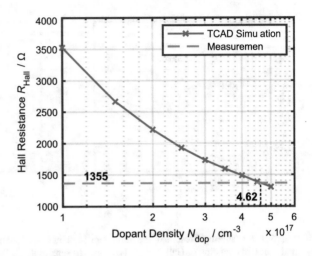

Table 5.3 Optimized parameter for TCAD simulation

Parameter	Symbol	Value
Dopant density P-substrate (Boron)	P_{dop}	$1 \cdot 10^{15} \mathrm{cm}^{-3}$
Dopant density N-well (Arsenic)	N_{dop}	$1 \cdot 10^{17} \mathrm{cm}^{-3} \Rightarrow 4.62 \cdot 10^{17} \mathrm{cm}^{-3}$
Thickness P-substrate (Boron)	t_{chip}	$5\,\mu\mathrm{m}$
Thickness N-well (Arsenic)	t_{Hall}	$1\,\mu\mathrm{m}$
Hall factor	r_{H}	$1.1 \Rightarrow 0.755$

sensitivity S_{A} additionally depends on the Hall factor r_{H} of the used technology.
Inserting Eq. 5.4 into 5.7 and solving for r_{H} results in

$$r_H = \frac{S_{\mathrm{A}} \cdot N_{\mathrm{dop}} \cdot t_{\mathrm{Hall}} \cdot q}{G \cdot I_{\mathrm{bias,H}}}. \tag{5.9}$$

S_{A} is the measured sensitivity for the corresponding bias current $I_{\mathrm{bias,H}}$, and
N_{dop} the adjusted dopant density of the previous R_{Hall} consideration ($N_{\mathrm{dop}} =
4.62 \cdot 10^{17} \mathrm{cm}^{-3}$). For measured sensitivity $S_{\mathrm{A}} = 10.1\,\mathrm{mV/T}$ with a bias current
of $I_{\mathrm{bias,H}} = 1\,\mathrm{mA}$ the Hall factor r_{H} results in 0.755 (instead of 1.1, as assumed in
the first simulations). Table 5.3 shows an overview of the new parameters for the
TCAD simulation of the Hall device. By adjusting the dopant density N_{dop} of the
active N-well and the Hall factor r_{H} of the used technology, the simulations agree
well with measurements, as depicted in Table 5.4. To further increase the absolute
sensitivity S_{A}, an additional P+ layer can be added at the surface of the active N-
well of the Hall device [10]. Hence, the effective thickness t_{Hall} of the Hall device
decreases, leading to 12 % higher S_{A}, as shown in Fig. 5.10.

For on-chip current sensing, the proposed Hall device with the additional P+
layer is placed besides a power line that carries the signal current to be measured
with a distance of 130 μm from its center and the center of the Hall device, as

Table 5.4 Comparison between optimized TCAD simulation and measurement results of the implemented cross-shaped Hall device

Parameter	Symbol	Measurement	Optimized TCAD simulation
Sensitivity @ $I_{bias,H} = 1\,mA$	S_A	10.1 mV/T	10.9 mV/T
Resistance	R_{Hall}	1355 Ω	1365 Ω

Fig. 5.10 Measured absolute sensitivity S_A of the Hall device with an additional P+ layer and $I_{bias,H} = 1\,mA$

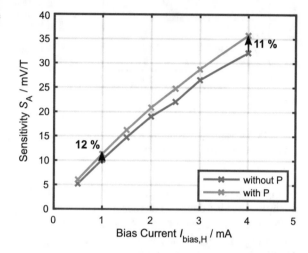

Fig. 5.11 Photograph of the implemented cross-shaped Hall device for on-chip current sensing

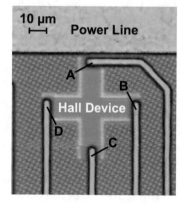

partially shown in Fig. 5.11. This corresponds to a geometric mean distance GMD of 125 μm, which results in a magnetic field B_z of 1.6 mT for $I_{meas} = 1\,A$. With this placement, the absolute sensitivity $S_A = 11.3\,mV/T$ at a bias current $I_{bias,H} = 1\,mA$ and the assumption of a homogeneous magnetic field $\vec{B}$, the current related sensitivity $S_{A,Imeas}$ of the implemented Hall device is defined as

$$S_{A,Imeas} = \frac{V_{Hall}}{I_{meas}} = S_A \cdot \frac{B_z}{I_{meas}} = 11.3\,\frac{mV}{T} \cdot \frac{1.6\,mT}{1\,A} = 18.08\,\frac{\mu V}{A}. \tag{5.10}$$

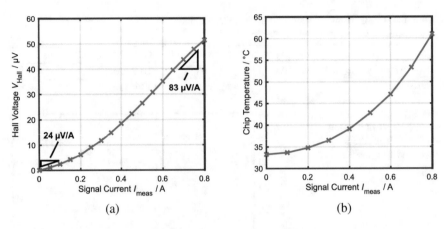

Fig. 5.12 Measurement results of the on-chip current sensing with $I_{bias,H} = 1$ mA. (**a**) measured Hall voltage V_{Hall} and (**b**) chip temperature

Figure 5.12a shows the measurement results of the on-chip current sensing with the proposed Hall device. The measured Hall voltage V_{Hall} does not increase linearly with increasing signal current I_{meas}, as shown in Fig. 5.12a. This results from the increasing chip temperature due to self-heating in the power line, as shown in Fig. 5.12b. The spot with the highest temperature of the chip is measured with the FLIR infrared camera E64501. Only for currents $I_{meas} \leq 0.1$ A, the current related Hall voltage is in the range as expected from Eq. 5.10.

As shown by the measurement results in Fig. 5.12 and published in [11, 12], the sensitivity S_A has a strong temperature dependence. This originates from the varying Hall factor r_H [3] and the changing mobility in the semiconductor [13, 14].

There are several methods to compensate this temperature dependence by generating a reference field for trimming or by calibration at specific temperatures. In [15, 16] a magnetic field $\vec{B}$ is generated with an integrated coil. In this case, the sensitivity S_A can be regulated by adjusting the bias current $I_{bias,H}$ of the Hall device.

A calibration technique is used in [17]. The Hall voltage V_{Hall} is measured at two specific temperatures and the temperature coefficient is calculated. For this purpose the temperature of the chip must be measured by a thermal sensing circuit. Even commercial product implementations are available for this purpose [18–20]. The compensation of the thermal drift was not further focused in this book.

As a conclusion, with a symmetrical cross-shaped Hall device the highest absolute sensitivity S_A can be achieved. In a first approximation, S_A only depends on L_{Hall}/W_{Hall} ratio and not on the absolute size, which is confirmed by TCAD simulations. The dopant density N_{dop} of the active N-well and the Hall factor r_H were determined from the measurement results of the implemented Hall device, which results in a good agreement of S_A between TCAD simulations and measurements. Measurements show a current related sensitivity $S_{A,Imeas}$ of $24\,\mu$V/A for currents to be measured $I_{meas} \leq 0.1$ A. At higher currents I_{meas}, the temperature of the chip increases, which leads to a sensitivity up to $83\,\mu V/A$. The linearity of the Hall voltage V_{Hall} depending on the magnetic field $\vec{B}$ and the bias current $I_{bias,H}$ are confirmed by measurement results.

5.3 Hall Device for Off-Chip Current Sensing

Off-chip current sensing is essential for high-power applications to provide the required galvanic isolation as derived in Sect. 2.1.2. To get a high sensitivity for DC and low-frequency current sensing, the chip with the vertical Hall device is intended to be mounted on top of a power module or a PCB and senses the magnetic field $\vec{B}$ in y-direction [2].

In the following, the design and parameters such as the sensitivity of the vertical Hall device are considered. Measurement results of the off-chip current sensing are discussed.

The sensitivity of a Hall device is the change of the Hall voltage V_{Hall} related to the change of the external magnetic field $\vec{B}$ (as for on-chip current sensing in Sect. 5.2). It depends on the parameters of the used CMOS technology, e.g., dopant density and thickness of the active N-well.

A basic variant of the vertical Hall device consists of the N+ terminals A to E, as shown in Fig. 5.13. These terminals are located in a N-well and placed in parallel to the magnetic field to be measured. The Hall device is biased by a bias current $I_{bias,H}$ at terminal A, which splits in the N-well to terminal B and C. In presence of any magnetic field $\vec{B}$, the current in the N-well flows asymmetrically, which results in a Hall voltage $V_{Hall} \neq 0\,V$ between terminals D and E [21, 22]. Similar to the planar Hall device, V_{Hall} is proportional to the magnetic field $\vec{B}$ and to $I_{bias,H}$.

Vertical Hall devices usually have a lower sensitivity S_A compared to planar Hall devices. To overcome this effect, several of these basic structures or variants

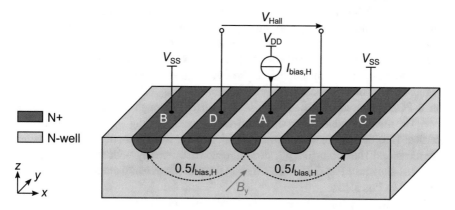

Fig. 5.13 Cross-section of a vertical Hall device

Fig. 5.14 Photograph of the
bare-chip with the vertical
Hall device mounted on top a
PCB for off-chip current
sensing

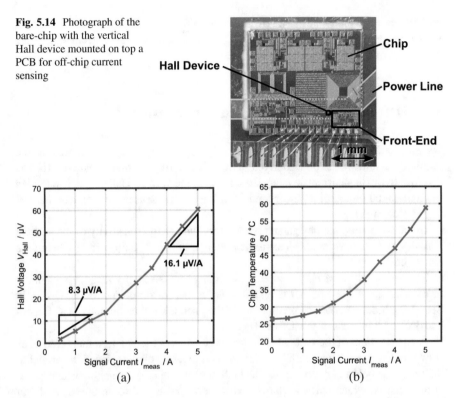

Fig. 5.15 Measurement results of off-chip current sensing with $I_{bias,H} = 16\,mA$: (**a**) measured
Hall voltage V_{Hall} and (**b**) chip temperature

can be combined in different ways to increase S_A [22–24]. Furthermore, due to
an optimized placement of these structures, multi-dimensional measurement of the
magnetic field is also possible [25–27].

Because of the well-defined orientation of the magnetic field $\vec{B}$ of the signal
current I_{meas}, in this work, only one basic structure of the vertical Hall device is
implemented in a 180 nm HV CMOS process. To improve the coupling between
the power line and the vertical Hall device, the chip was back-grinded to a
total thickness of 60 μm, similar to the helix-shaped Rogowski coil in Sect. 3.3.
Figure 5.14 shows a photograph of the bare-die chip, which is mounted directly
on a PCB for contactless sensing of the current through the underlying power
line. The measurement results of the off-chip current sensing with the implemented
vertical Hall device are shown in Fig. 5.15. Similar to the on-chip current sensing,
the measured Hall voltage V_{Hall} does not increase linearly with the signal current
I_{meas}, as shown in Fig. 5.15a. This results from the increasing chip temperature, as
depicted in Fig. 5.15b.

Due to the chip thickness of 60 μm, the distance between the Hall device and the
power line is minimal. By considering the thickness of the power line of 35 μm and

an insulation varnish of approximately 25 μm the geometric mean distance (GMD) between the power line and the vertical Hall device is about 88 μm. This corresponds to a current related magnetic field of 2.3 mT/A. For currents $I_{meas} < 2$ A the resulting absolute sensitivity can be approximated by

$$S_A = \frac{S_{A,Imeas}}{\dfrac{B_z}{I_{meas}}} \approx \frac{8.3\,\mu V/A}{\dfrac{2.3\,mT}{1\,A}} = 3.6\,\frac{mV}{T}. \tag{5.11}$$

As expected, the measured absolute sensitivity $S_A = 3.6$ mV/T is lower than the measured sensitivity of the planar Hall device ($S_A = 10.1$ mV/T) due to the shallow depth of the N-well.

In conclusion, the vertical Hall device was implemented in a 180 nm HV CMOS technology. Measurements show that with a bias current of $I_{bias,H} = 16$ mA an acceptable sensitivity of 8.3 μV/A can be achieved for off-chip current sensing by a vertical Hall device.

5.4 Summary

A Hall sensor is well suitable for sensing DC and low-frequency signal currents. In this work, Hall sensors are used to extend the limited low-frequency range of a Rogowski coil towards DC signal currents. Taking into account the geometry of a Hall device, a cross-shaped Hall device is optimized for on-chip current sensing. A vertical Hall device is implemented for off-chip current sensing.

With an optimized design of a cross-shaped Hall device for on-chip current sensing, a sensitivity $S_{A,Imeas}$ of 24 μV/A at a bias current $I_{bias,H}$ of 1 mA is achieved. The Hall device is placed besides a power line that carries the signal current to be measured with a distance of 130 μm from its center and the center of the Hall device. A galvanically isolation for integrated current sensing of high-voltage switching applications is provided. Measurement results are used for extraction of the dopant density N_{dop} of the active N-well and the Hall factor r_H. This allows an accurate simulation and optimization of the Hall device by TCAD simulations. The measured sensitivity S_A of 10.1 mV/T is confirmed by a TCAD simulation.

For off-chip current sensing, the vertical Hall device achieves a sensitivity $S_{A,Imeas}$ of 8.3 μV/A at a bias current $I_{bias,H}$ of 16 mA. The distance between Hall device and the power line is optimized by back-grinded chips with a total thickness of 60 μm in comparison to the regular thickness of 250 μm. The implementation of a vertical Hall device allows to place the chip on top of a current carrying conductor and to measure the current flowing under it with galvanic isolation.

Table 5.5 Calculation of geometry factor G of different geometries of symmetrical Hall devices

Shape	Geometry factor G	$e/d = f(W_{Hall}/L_{Hall})$
Cross	$G = 1 - 1.045 \cdot e^{(-\pi \cdot L_{Hall}/W_{Hall})}$	–
Octagon	$G = 1 - 1.94 \cdot \left(\dfrac{e/d}{1 + 0.0414 \cdot e/b} \right)^2$	$e/d = \dfrac{W_{Hall}/L_{Hall}}{W_{Hall}/L_{Hall} + \sqrt{2}}$
Square 1	$G = 1 - 1.062 \cdot e/b$	$e/d = \dfrac{W_{Hall}/L_{Hall}}{W_{Hall}/L_{Hall} + 2}$
Square 2	$G = 1 - 0.696 \cdot (e/d)^2$	$e/d = \dfrac{2}{2 + W_{Hall}/L_{Hall}}$

Appendix

Geometry Factor of Symmetrical Hall Devices

Table 5.5 lists the calculated geometry factors G for different geometries. G of the cross-shape is calculated from the dimensions L_{Hall} and W_{Hall}. While the calculation of G of the octagon-, square 1-, and square 2-shape is done with the auxiliary variables d and e. The variables e is the total length of all contacts and the variables d is the total length of all edges and thus the circumference of the respective shape. In order to be able to compare these shapes with the cross-shape Hall device, a conversion is introduced. Therefore, e.g., the cross is inserted into the octagon and e/d is calculated according to Eq. 5.12 depending on W_{Hall} and L_{Hall}. With this conversion method, the geometry factor G can be calculated for the different shapes and compared by each other.

$$\frac{e}{d} = \frac{\sum \text{contact length}}{\text{circumference}} \tag{5.12}$$

References

1. Funk, T., & Wicht, B. (2018). A fully integrated DC to 75 MHz current sensing circuit with on-chip Rogowski coil. In *Proceedings of the IEEE Custom Integrated Circuits Conference (CICC)* (pp. 1–4). https://doi.org/10.1109/CICC.2018.8357028.
2. Funk, T., Groeger, J., & Wicht, B. (2019). An integrated and galvanically isolated DC-to-15.3 MHz hybrid current sensor. In *Proceedings of the IEEE Applied Power Electronics Conference and Exposition (APEC)* (pp. 1010–1013).
3. Popovic, R. S. (2004). *Hall effect devices* (2nd New ed.). Milton Park: Taylor & Francis Ltd. ISBN: 0750308559.
4. Crescentini, M., Biondi, M., Romani, A., Tartagni, M., & Sangiorgi, E. (2017). Optimum design rules for CMOS hall sensors. *Sensors, 17*(4), 765.
5. Demierre, M. (2003). Improvements of CMOS Hall microsystems and application for absolute angular position measurements. Ph.D. Thesis. EPF Lausanne.

6. Paun, M.-A., Sallese, J.-M., & Kayal, M. (2013a). Hall effect sensors design, integration and behavior analysis. *Journal of Sensor and Actuator Networks, 2*, 85–97.
7. Paun, M.-A., Sallese, J.-M., & Kayal, M. (2013b). Comparative study on the performance of five different hall effect devices. *Sensors, 13*(2), 2093–2112.
8. Schwarzbeck (Rev. B). Circular Helmholtz coils. In *Datasheet HHS 5201-98.*
9. Xu, Y., & Pan, H.-B. (2011). An improved equivalent simulation model for CMOS integrated Hall plates. *Sensors, 11*(6), 6284–6296. ISSN: 1424-8220. http://www.ncbi.nlm.nih.gov/pmc/articles/PMC3231436/.
10. Fan, Q., Huijsing, J. H., & Makinwa, K. A. A. (2012). A capacitively-coupled chopper operational amplifier with 3μV offset and outside-the-rail capability. In *Proceedings of the ESSCIRC (ESSCIRC) 2012* (pp. 73–76). https://doi.org/10.1109/ESSCIRC.2012.6341259.
11. Cholakova, I. N., Takov, T. B., Tsankov, R. T., & Simonne, N. (2012). Temperature influence on Hall effect sensors characteristics. In *Proceedings of the 20th Telecommunications Forum (TELFOR)* (pp. 967–970). https://doi.org/10.1109/TELFOR.2012.6419370.
12. Feng, W., Dong, M., & Zhangsui, X. (2007). A research about the temperature compensation of Hall sensor. In *Proceedings of the 8th International Conference Electronic Measurement and Instruments* (pp. 4-131-4-134). https://doi.org/10.1109/ICEMI.2007.4351100.
13. Paun, M., Sallese, J., & Kayal, M. (2012). Temperature considerations on Hall effect sensors current-related sensitivity behaviour. In *Proceedings of the 19th IEEE International Conference Electronics, Circuits and Systems (ICECS 2012)* (pp. 201–204). https://doi.org/10.1109/ICECS.2012.6463766.
14. Manic, D., Petr, J., & Popovic, R. S. (2000). Temperature cross-sensitivity of Hall plate in submicron CMOS technology. In *Sensors and Actuators A: Physical, 85*(1), 244–248. ISSN: 0924-4247. http://www.sciencedirect.com/science/article/pii/S092442470000399X.
15. Jiang, J., & Makinwa, K. (2016). A hybrid multi-path CMOS magnetic sensor with 76 ppm/ °C sensitivity drift. In *Proceedings of the ESSCIRC Conference 2016: 42nd European Solid-State Circuits Conference* (pp. 397–400). https://doi.org/10.1109/ESSCIRC.2016.7598325.
16. Jiang, J., & Makinwa, K. A. A. (2017). A hybrid multi-path CMOS magnetic sensor with 76 ppm/ °C sensitivity drift and discrete-time ripple reduction loops. In *IEEE Journal of Solid-State Circuits, 52*(7), 1876–1884. ISSN: 0018-9200. https://doi.org/10.1109/JSSC.2017.2685462.
17. Blanchard, H., De Raad Iseli, C., & Popovic, R. S. (May 1997). Compensation of the temperature-dependent offset drift of a Hall sensor. *Sensors and Actuators A: Physical, 60*(1), 10–13. ISSN: 0924-4247. http://www.sciencedirect.com/science/article/pii/S0924424796014112.
18. Innosen (Rev2.0. 4.140108). Temperature compensation bipolar Hall effect sensor. In *Datasheet ES413/ES513.*
19. Micronas (Edition Dec. 8, 2008). Linear Hall-effect sensor IC. In *Datasheet HAL 411.*
20. Instruments, Texas (2018). Automotive ratiometric linear Hall effect sensor. In *Datasheet DRV5055-Q1.*
21. Phetchakul, T., Poonsawat, S., & Poyai, A. (2016). The effect of deviation current to 5-contacts vertical Hall device. In *Proceedings of the Telecommunications and Information Technology (ECTI-CON) 2016 13th International Conference Electrical Engineering/Electronics, Computer* (pp. 1–4). https://doi.org/10.1109/ECTICon.2016.7561375.
22. Kaufmann, T., Purkl, F., Ruther, P., & Paul, O. (2011). Novel coupling concept for five-contact vertical Hall devices. In *Proceedings of the Actuators and Microsystems Conference 2011 16th International Solid-State Sensors* (pp. 2855–2858). https://doi.org/10.1109/TRANSDUCERS.2011.5969126.
23. Sung, G., Wang, W., & Yu, C. (2017). Analysis and modeling of one-dimensional folded vertical Hall sensor with readout circuit. *IEEE Sensors Journal, 17*(21), 6880–6887. ISSN: 1530-437X. https://doi.org/10.1109/JSEN.2017.2754295.
24. Heidari, H., Bonizzoni, E., Gatti, U., Maloberti, F., & Dahiya, R. (2015). Optimal geometry of CMOS voltage-mode and current-mode vertical magnetic Hall sensors. In *Proceedings of the IEEE SENSORS* (pp. 1–4). https://doi.org/10.1109/ICSENS.2015.7370365.

25. Sung, G.-M., Gunnam, L. C., Wang, H.-K., Lin, W.-S. (2018). Three-dimensional CMOS differential folded Hall sensor with bandgap reference and readout circuit. In *IEEE Sensors Journal, 18*(2), 517–527. ISSN: 1530-437X. https://doi.org/10.1109/JSEN.2017.2777485.
26. Paranjape, M., Ristic, L., & Filanovsky, I. (1991). A 3-D vertical hall magnetic field sensor in CMOS technology. In *Proceedings of the TRANSDUCERS '91: 1991 International Conference Solid-State Sensors and Actuators Digest of Technical Papers* (pp. 1081–1084). https://doi.org/10.1109/SENSOR.1991.149085.
27. Sander, C., Leube, C., & Paul, O. (2015). Novel compact two-dimensional CMOS vertical Hall sensor. In *Proceedings of the Actuators and Microsystems (TRANSDUCERS) 2015 Transducers - 2015 18th International Conference Solid-State Sensors* (pp. 1164–1167). https://doi.org/10.1109/TRANSDUCERS.2015.7181135.

Chapter 6
Hall Current Sensor Front-End

The output signal of a Hall device needs to be processed by a sensor front-end to achieve a high sensitivity $S_{out,H}$ for the subsequent signal evaluation, as indicated in Fig. 6.1. The output voltage V_{Hall} of a Hall device has only small amplitudes for current sensing, as shown in Sects. 5.2 and 5.3, while the DC operation point varies strongly depending on the bias current $I_{bias,H}$ of the Hall device. Therefore, a sensor front-end with a high gain and a DC-decoupled signal transmission is required. Furthermore, a bandwidth of $>20\,kHz$ is needed to extend the limited low-frequency range of the Rogowski path towards DC signal currents.

Modulation techniques of the Hall voltage V_{Hall} for off-chip current sensing are proposed in Sects. 6.1 and 6.2. Both enable DC-decoupled signal transmission and amplification in Sect. 6.2 [1, 2].

6.1 Signal Modulation for On-Chip Current Sensing

Low-frequency errors such as offset and noise limit the minimum signal current to be measured. Therefore, due to the multiple symmetrical layout of the cross-shaped Hall device, the proven four-phase current spinning technique according to [3, 4] is implemented. This allows offset reduction of the Hall device and in addition offset and noise reduction of the sensor front-end for on-chip current sensing.

Figure 6.2a shows the implemented block diagram including current spinning and the capacitively coupled sensor front-end [1]. Due to gradients in dopant density or temperature in the active area of a Hall device, an unbalanced Hall resistance R_{Hall} is generated, which leads to an offset voltage $V_{Hall,off}$. The current spinning technique is used to modulate the Hall voltage V_{Hall} to the high-frequency region. Therefore, the bias current $I_{bias,H}$ and the output voltage V_{Hall} rotate by 90° for each phase. Furthermore, the polarity in phase 2 and 4 is reversed when sensing the Hall

© The Editor(s) (if applicable) and The Author(s), under exclusive license to
Springer Nature Switzerland AG 2020
T. Funk, B. Wicht, *Integrated Wide-Bandwidth Current Sensing*,
https://doi.org/10.1007/978-3-030-53250-5_6

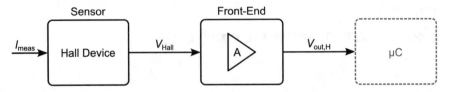

Fig. 6.1 Signal processing for current sensing with a Hall device

voltage V_{Hall}, as shown in Fig. 6.2b. As a result, V_{Hall} has a DC component from the offset voltage $V_{Hall,off}$ and an AC component from the signal voltage $V_{Hall,sig}$, which is generated by a magnetic field $\vec{B}$, as shown in Fig. 6.2c. This has the advantage that the offset $V_{Hall,off}$ is suppressed by the capacitively transmission to the subsequent sensor front-end. The unbalanced Hall resistance R_{Hall} is indicated by R_Δ in Fig. 6.2b.

The simplified circuit diagram of the implemented four-phase spinning technique is shown in Fig. 6.3. The switches S_1 to S_8 rotate the bias current $I_{bias,H}$ and the switches S_9 to S_{16} the output voltage V_{Hall} by 90° for each phase. For a bias current $I_{bias,H}$ of up to 4 mA and a measured Hall resistance of 1355Ω, the voltage across the Hall device reaches 5 V. Therefore, the switches are implemented using 5 V-devices, since these have the required dielectric strength against substrate (5.5 V). To reduce voltage peaks while switching between the individual phases, the switches are controlled with a non-overlapping clock. Figure 6.4 shows the 4-bit shift register used to generate spinning phases 1 to 4 and the logic used to generate the non-overlapping clock ϕ_1 to ϕ_8 for switch control. The non-overlap time ΔT is set to 5 ns by buffers stages. This allows for current spinning up to 10 MHz.

6.2   Signal Modulation for Off-Chip Current Sensing

In contrast to the cross-shaped Hall device, the layout of the implemented vertical Hall device has a single axial symmetry. Therefore, chopping is implemented for noise and offset reduction of the sensor front-end for off-chip current sensing [2].

Figure 6.5 shows the implemented block diagram for vertical Hall sensing. The chopper element CH_1 modulates the Hall voltage V_{Hall} into the high-frequency region, which is amplified by the subsequent capacitively coupled sensor front-end. With chopping, the noise and offset V_{OS} of the amplifier A_1 can be significantly reduced, while the offset voltage $V_{Hall,off}$ of the Hall sensor cannot be eliminated by this technique.

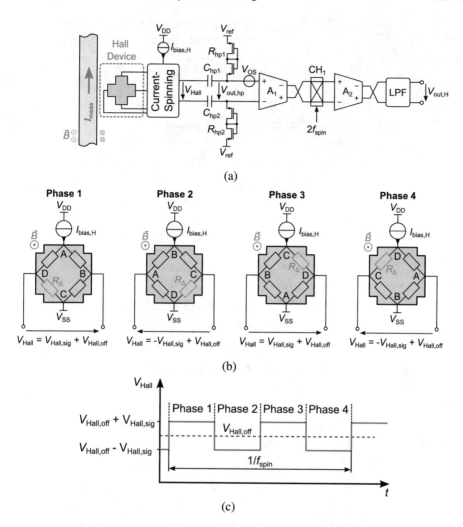

(a)

(b)

(c)

Fig. 6.2 Block diagram for (**a**) on-chip current sensing using a cross-shaped Hall device with current spinning (**b**) phases and (**c**) waveform for offset cancelation and frequency modulation

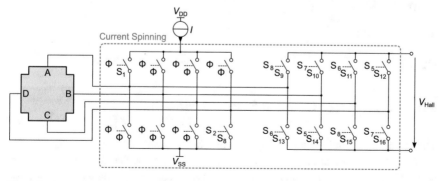

Fig. 6.3 Schematic of four-phase current spinning

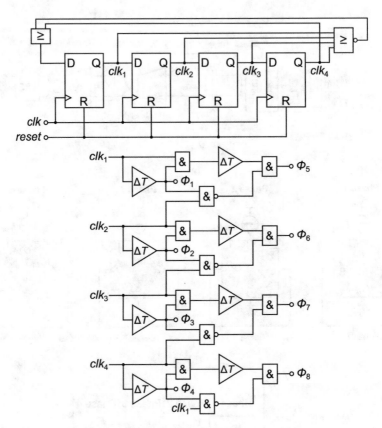

Fig. 6.4 Schematic of switch control block for four-phase current spinning

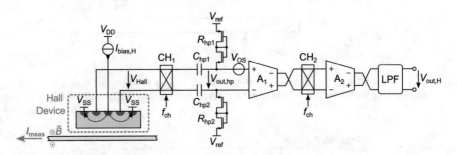

Fig. 6.5 Sensor front-end for off-chip current sensing with a vertical Hall device

6.3   Capacitively Coupled Sensor Front-End

The modulated Hall voltage V_{Hall} for on-chip current sensing as well for off-chip current sensing are amplified by a capacitively coupled sensor front-end. This section describes the implemented capacitively transmission using a high-pass filter with a pseudo-resistor, the amplification and the measurement results for on-chip and off-chip current sensing [1, 2].

Since the sensitivity S_A of the Hall device is set by the bias current $I_{bias,H}$, the DC operating point of the Hall voltage V_{Hall} varies. This variation can be suppressed by capacitively transmission of V_{Hall} using a high-pass filter with $C_{hp1/2}$ and $R_{hp1/2}$ and allows amplification of V_{Hall} independent of $I_{bias,H}$. To reduce the noise of the sensor front-end at the same time, a modulation frequency >1 MHz is chosen, since the corner frequency f_c between the $1/f$ noise and the thermal noise of the implemented amplifiers are in the MHz region, as described in Sect. 4. After capacitively transmission, the modulated voltage V_{Hall} is amplified by the symmetrical amplifiers A_1 and A_2 to achieve high sensitivity, as shown in Figs. 6.2a and 6.5. These amplifiers are both implemented according to Appendix. The chopper element between A_1 and A_2 is used to demodulate the amplified Hall voltage back to DC and to modulate the offset voltage V_{OS} and noise of A_1 into the high-frequency region. After the second amplification by amplifier A_2, the resulting ripple and the modulated noise is filtered by a low-pass filter (LPF).

For capacitively transmission, the high-pass filter must be designed to transmit the modulated Hall voltage V_{Hall} without attenuation. The cut-off frequency f_{hp} of the filter must therefore be significantly lower than the modulation frequency. In addition, the capacitance C_{hp} must be chosen to prevent a capacitive voltage divider with the parasitic input capacitance of the amplifier A_1 [5]. By choosing $C_{hp} = 20$ pF the capacitive voltage divider can be neglected and maximum amplification can be ensured. Therefore, R_{hp} below 1 MΩ would be sufficient to achieve the required cut-off frequency $f_{hp} < 1$ MHz and could be implemented by ploy resistors. However, the voltage $V_{out,hp}$ must be constant during the individual phases of current spinning and chopping and may not be discharged via R_{hp}. This means that a significantly higher resistance value is required. In order to cover lower modulation frequencies down to 100 kHz, $V_{out,hp}$ must be kept constant for 5 µs. This requires a resistor $R_{hp} > 1$ GΩ.

For the implementation of a resistor with $R_{hp} > 1$ GΩ an area-efficient pseudo-resistor consisting of two PMOS transistors is proposed. Each PMOS transistor is operated in a diode configuration and the respective body diode is used for symmetrical operation, as shown in Fig. 6.6a. Due to the series configuration of the two transistors, the current I_{Rhp} and thus the resistance R_{hp} is the same in both directions, as indicated in Fig. 6.6b.

This implementation achieves a resistance value of 7.5 GΩ with an area of only 84 µm^2. As a result, the cut-off frequency f_{hp} of the high-pass filter is shifted to approximately 1 Hz. In this case, only the startup behavior of the sensor front-end is limited by the large time constant, while this is improved by a short circuit of R_{hp} during startup.

After the capacitively transmission, the Hall voltage V_{Hall} is amplified by the two amplifiers A_1 and A_2 with a DC gain of 40 dB and a bandwidth of 26 MHz for each amplifier. Due to the high modulation frequency of 4.5 MHz a low output noise of 4.6 mVrms can be reached. An offset voltage of 233 mV occurs at the output $V_{out,H}$

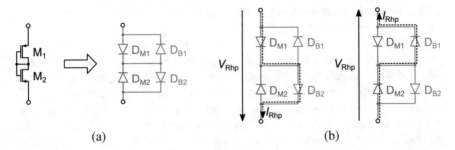

Fig. 6.6 Pseudo-resistor: PMOS transistors in (**a**) diode configuration and (**b**) current flow consideration

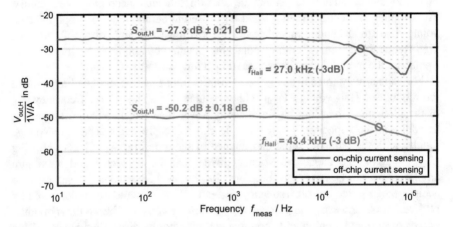

Fig. 6.7 Measured transfer behavior of on-chip current sensing with a cross-shaped Hall device and off-chip current sensing with a vertical Hall device

of the sensor front-end. This offset voltage is caused by the input offset of A_2, and could be reduced by a compensation loop according [3].

The measured transfer behavior of on-chip current sensing and off-chip current sensing is shown in Fig. 6.7 [1, 2]. As for the measurement of the transfer behavior of the Rogowski coil, signal currents I_{meas} with an amplitude of 0.5 A for the on-chip current sensing and 1.4 A for the off-chip current sensing are generated. For a bias current of $I_{bias,H} = 3.3$ mA the cross-shaped Hall device for on-chip current sensing has a bandwidth f_{Hall} of 27 kHz and a sensitivity $S_{out,H}$ of 43.3 mV/A. The measured sensitivity for this $I_{bias,H}$ is slightly lower than expected from Sect. 5.3 due to the deviation of gain of A_1 and A_2 and the influence of the capacitive voltage divider between the high-pass filter and the input capacitance of A_1. However, $I_{bias,H}$ can be increased up to 4 mA and the lower gain of the front-end can be compensated. The vertical Hall device and the implemented sensor front-end achieves a bandwidth f_{Hall} of 43 kHz and a sensitivity $S_{out,H}$ of 3.1 mV/A for a bias current $I_{bias,H}$ of 16 mA.

6.4 Summary

A capacitively coupled sensor front-end is utilized for both on-chip current sensing and off-chip current sensing. Four-phase current spinning technique is applied for on-chip current sensing. Thereby, the Hall voltage V_{Hall} is modulated by a spinning frequency. This way, the offset voltage of a Hall device, caused by gradients in doping density or temperature, is separated from the signal voltage. Furthermore, it allows to reduce the noise and the offset of the following sensor front-end. For off-chip current sensing, chopping is implemented to reduce the noise and the offset of the sensor front-end. With a modulation frequency of 4.5 MHz a low output noise of 4.6 mVrms and an output offset voltage of 233 mV is reached.

For a bias current $I_{bias,H} = 3.3$ mA of the cross-shaped Hall device, a bandwidth of 27 kHz and sensitivity of 43.3 mV/A is achieved for on-chip current sensing. The vertical Hall sensing reaches a bandwidth of 43 kHz and a sensitivity of 3.1 mV/A for a bias current of 16 mA. Both sensitivities correspond to the sensitivity of the respective current sensing with the Rogowski coil for on-chip and off-chip current sensing. The achieved bandwidths are sufficient to extend the limited low-frequency range of the Rogowski coil towards DC signal currents.

Appendix

Fully Differential Symmetrical Amplifier

The implementation of the fully differential symmetrical amplifier for signal processing is shown in Fig. 6.8. For high bandwidth optimization, the current mirrors M_5 to M_7 and M_4 to M_6 are implemented with a ratio $n = 5$. Furthermore,

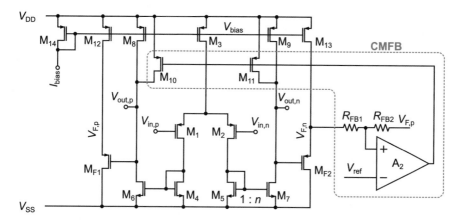

Fig. 6.8 Schematic of a fully differential symmetrical amplifier

the output stage is optimized for low parasitic capacitances. Therefore, the source-follower M_{F1} and M_{F2} buffers the output of A_2 for the required Common-mode feedback (CMFB). This enables a nominal DC gain of 40 dB with a bandwidth of 26 MHz.

References

1. Funk, T., & Wicht, B. (2018). A fully integrated DC to 75 MHz current sensing circuit with on-chip Rogowski coil. In *Proceedings of the IEEE Custom Integrated Circuits Conference (CICC)* (pp. 1–4). https://doi.org/10.1109/CICC.2018.8357028.
2. Funk, T., Groeger, J., & Wicht, B. (2019). An integrated and galvanically isolated DC-to-15.3 MHz hybrid current sensor. In *Proceedings of the IEEE Applied Power Electronics Conference and Exposition (APEC)* (pp. 1010–1013).
3. Jiang, J., & Makinwa, K. A. A. (2015). A multi-path CMOS hall sensor with integrated ripple reduction loops. In *Proceedings of the IEEE Asian Solid-State Circuits Conference (A-SSCC)* (pp. 1–4). https://doi.org/10.1109/ASSCC.2015.7387504.
4. Jiang, J., Kindt, W. J., & Makinwa, K. A. A. (2014). A continuous-time ripple reduction technique for spinning-current hall sensors. *IEEE Journal of Solid-State Circuits, 49*(7), 1525–1534. ISSN: 0018-9200. https://doi.org/10.1109/JSSC.2014.2319252.
5. Fan, Q., Huijsing, J. H., & Makinwa, K. A. A. (2012). A capacitively-coupled chopper operational amplifier with 3μV offset and outside-the-rail capability. In *Proceedings of the ESSCIRC (ESSCIRC) 2012* (pp. 73–76). https://doi.org/10.1109/ESSCIRC.2012.6341259.

Chapter 7
Hybrid Current Sensor

A hybrid current sensor, consisting of a Hall sensor and a Rogowski coil, is well suitable for wide-bandwidth current sensing [1–3]. In this work, the Hall paths for on-chip and off-chip current sensing from Chap. 5 can be connected to both the open-loop and the closed-loop Rogowski paths for on-chip and off-chip current sensing from Sects. 4.1 and 4.2. In this chapter, a connection network for on-chip current sensing with the cross-shaped Hall sensor and the planar Rogowski coil with the open-loop sensor front-end is explored. By combining these two sensor principles, signal currents from DC up to 71 MHz can be measured.

In the previous Chaps. 3–6 the DC and low-frequency current sensing with a Hall sensor and the high-frequency current sensing with a Rogowski coil are introduced. Both sensor principles are designed to extend the frequency range with the other sensor principle. The simplified block diagram of a hybrid current sensor is shown in Fig. 7.1. As the Rogowski path is used to measure AC signal currents, the output voltage $V_{out,R}$ can be capacitively transmitted to the output of the hybrid current sensor [4]. The DC operation point of the hybrid current sensor is set by the Hall path. This allows the connection of both sensor principles by a passive high-pass filter.

The high-pass filter with a cut-off frequency of 1.6 kHz consist of the resistance $R_{hp1/2} = 300\,k\Omega$ and the capacitance $C_{hp1/2} = 330\,pF$. The passive filter implementation ensures that the output voltage $V_{out,R}$ of the Rogowski path is transmitted to the output of the hybrid current sensor undamped. The value of $R_{hp1/2}$ must be chosen low, as it forms a voltage divider with the input resistance of the voltage probe ($R_{in,Probe} = 1\,M\Omega$) in the experimental setup. In a future commercial implementation, this voltage divider could be avoided by using a buffer with low input resistance. At the same time, however, $C_{hp1/2}$ cannot be increased arbitrarily, as this increases the capacitive load at the output of the Rogowski path and will reduce the bandwidth.

Figure 7.2 shows the measured large signal transfer behavior of the on-chip current sensing of the proposed hybrid current sensor along with the curves of

T. Funk, B. Wicht, *Integrated Wide-Bandwidth Current Sensing*, https://doi.org/10.1007/978-3-030-53250-5_7

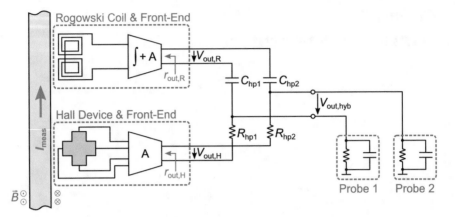

Fig. 7.1 Block diagram of a hybrid current sensor for wide-bandwidth current sensing

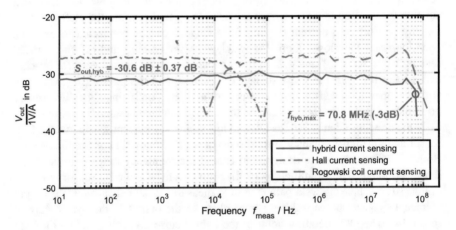

Fig. 7.2 Measured transfer behavior of on-chip current sensing with a hybrid current sensor

the Hall path as well as the Rogowski path from Sects. 4.1 and 6.3, respectively. Sinusoidal signal currents I_{meas} with an amplitude of 0.5 A are generated by a 4-quadrant amplifier (DC–100 kHz), followed by an RF broadband power amplifier (10 kHz–100 MHz). With the hybrid current sensor, a wide frequency range from DC to 71 MHz with a sensitivity $S_{\mathrm{out,hyb}}$ of -30.6 dB (29.5 mV/A) can be covered. The measured sensitivity $S_{\mathrm{out,hyb}}$ of the hybrid current sensor is approximately 3 dB lower than the measured sensitivity of the Rogowski path ($S_{\mathrm{out,R}} = -27.2$ dB). This is because of the higher bias current in the output stage of the Rogowski path and the resulting higher output resistance $r_{\mathrm{out,R}}$. The bias current of the output stage of the Rogowski path has to be increased due to the high capacitive load of the additional high-pass filter. This relationship is confirmed by a simulation.

For a constant sensitivity $S_{\mathrm{out,hyb}}$ of the hybrid current sensor, the sensitivity $S_{\mathrm{out,H}}$ of the Hall path is adjusted to the sensitivity $S_{\mathrm{out,R}}$ of the Rogowski path by setting the bias current $I_{bias,H}$ to 6 mA. The bias current of the Hall path in the hybrid current sensor is significantly higher than in the measurement in Sect. 6. This results from the voltage divider between the high-pass resistance R_{hp} and the input resistance of the voltage probe. Due to the high spinning frequency $f_{spin} = 8.5$ MHz of the Hall path, a chopping frequency of $f_{ch} = 9$ MHz of the Rogowski path and the large high-pass capacitance $C_{hp1/2}$ at the output of the hybrid sensor, a relatively low output noise of 1.4 mVrms is achieved. This is lower compared to the noise voltages of the individual sensing paths (2.2 mVrms and 4.6 mVrms), since the connection network additionally reduces the output noise. The low output offset of 7 mV is favored by the high spinning frequency f_{spin}, but can also be attributed to statistical fluctuations. In Sect. 6.3 the same sensor front-end of the Hall path has a much higher offset voltage of 233 mV at the output.

In conclusion, the previously introduced current sensing principles can be connected by a high-pass filter, forming a hybrid current sensor for on-chip current sensing with a high bandwidth from DC up to 71 MHz. This exceeds the bandwidth of fully integrated state-of-the-art hybrid current sensors by factor of 23 [4] and discrete state-of-the-art hybrid current sensors by factor of 1.4 [5]. Due to the high spinning and chopping frequency and the additional high-pass filter, a relatively low output noise voltage of 1.4 mVrms is achieved.

References

1. Dalessandro, L., Karrer, N., & Kolar, J. W. (2007). High-performance planar isolated current sensor for power electronics applications. *IEEE Transactions on Power Electronics, 22*(5), 1682–1692. ISSN: 0885-8993. https://doi.org/10.1109/TPEL.2007.904198.
2. Karrer, N., Hofer-Noser, P., & Henrard, D. (1999). HOKA: a new isolated current measuring principle and its features. In *Proceedings of the Conference Record of the 1999 IEEE Industry Applications Conference. Thirty-Forth IAS Annual Meeting (Cat. No.99CH36370)* (Vol. 3, pp. 2121–2128). https://doi.org/10.1109/IAS.1999.806028.
3. Karrer, N., & Hofer-Noser, P. (1999). A new current measuring principle for power electronic applications. In *Proceedings of the (Cat. No.99CH36312) 11th International Symposium Power Semiconductor Devices and ICs. ISPSD'99* (pp. 279–282). https://doi.org/10.1109/ISPSD.1999. 764117.
4. Jiang, J., & Makinwa, K. A. A. (2017). Multipath wide-bandwidth CMOS magnetic sensors. *IEEE Journal of Solid-State Circuits, 52*(1), pp. 198–209. ISSN: 0018-9200. https://doi.org/10. 1109/JSSC.2016.2619711.
5. Tröster, N., Wölfle, J., Ruthardt, J., & Roth-Stielow, J. (2017). High bandwidth current sensor with a low insertion inductance based on the HOKA principle. In *Proceedings of the 19th European Conference Power Electronics and Applications (EPE'17 ECCE Europe)* (pp. P.1– P.9). https://doi.org/10.23919/EPE17ECCEEurope.2017.8099003.

Chapter 8
Conclusion

An increasing need for wide-bandwidth current sensing for monitoring and control is observed in several applications and systems like power generation, e-mobility and automotive, smart grid, automation, and Internet of Things. This is mainly driven by the trend towards higher integration of power electronic circuits. The increasing switching frequency of highly integrated voltage converters, higher supply voltages, and the use of power switches such as Insulated-gate bipolar transistor (IGBT) or semiconductors like Gallium nitride (GaN) and Silicon carbide (SiC) are examples, related to this trend.

This book focuses on wide-bandwidth current sensing for internal signal currents (on-chip current sensing) and external signal currents (off-chip current sensing) by hybrid current sensor. The combination of a Hall sensor and a fully integrated Rogowski coil in one chip allows to measure DC currents and fast current changes with a high bandwidth. A planar Rogowski coil and a cross-shaped Hall device for on-chip current sensing and a helix-shaped Rogowski coil and a vertical Hall device for off-chip current sensing are implemented. An on-chip current sensor with a bandwidth of >50 MHz is required to reproduce the signal current of a voltage converter operating at a switching frequency in the multi-MHz region. For power electronic applications, like high-voltage converters or motor drivers, an off-chip current sensor with a bandwidth of >10 MHz is required to reproduce fast current transients. Often galvanic isolation is required to handle the high voltages. A study on state-of-the-art current sensors shows that current sensors are available for on-chip and off-chip current sensing, but they do not reach the required bandwidth in the multi-MHz range and/or they show several obstacles like high cost and large size.

The operation principle of a Rogowski coil is used to measure fast changing signal currents. For on-chip current sensing, a planar coil with an area of 0.35 mm^2 is placed besides a power line for the first time in this work. It measures internal signal currents by means of the magnetic field, associated with the signal current. A model

T. Funk, B. Wicht, *Integrated Wide-Bandwidth Current Sensing*, https://doi.org/10.1007/978-3-030-53250-5_8

of a fully integrated planar Rogowski coil allows to determine the best trade-off between bandwidth and sensitivity for the design of a planar coil. Ansys HFSS 3-D field simulation confirms that with an optimized design and an optimized layout a bandwidth of 1.5 GHz with a sufficient sensitivity is achieved. The conventional equivalent circuit of a Rogowski coil turns out to be not sufficient for planar on-chip current sensing and must be extended. By taking into account the coupling capacitances between the power line and the coil, the immunity against capacitive voltage coupling is improved by a factor of 100. Current sensing with superior characteristics is possible. For the first time, a helix-shaped coil, with the lower and upper metal layers of a CMOS technology, is implemented for off-chip current sensing. This offers the opportunity to mount the chip directly on top of a power line or power module and to measure the current flowing underneath. A dedicated model is derived, which contains all parasitic components of a helix-shaped Rogowski coil. This allows to determine the best trade-off between bandwidth and sensitivity for the design of a helix-shaped coil for off-chip current sensing. A helix-shaped coil with an area 0.75 mm^2 achieves a bandwidth of 200 MHz. Measurements confirm that the sensitivity of the Rogowski coil can be doubled by back-grinded chips with a total thickness of 60 μm compared to the regular thickness of 250 μm.

An active integrator circuit for open-loop current sensing is investigated as a sensing front-end for the Rogowski coil. Due to the combination of two active integrator stages a wide frequency range is covered by the Rogowski path. Experimental results confirm that frequency components 15 kHz $< f_{meas} <$ 75 MHz can be measured by the implemented planar Rogowski coil with a sensitivity of 43.7 mV/A. Likewise, frequency components 16 kHz $< f_{meas} <$ 15 MHz can be measured by the implemented helix-shaped Rogowski coil with a sensitivity of 3.1 mV/A. The bandwidth of 75 MHz (planar Rogowski coil) and 15 MHz (helix-shaped Rogowski coil) exceeds the state-of-the-art current sensing with a fully integrated coil by a factor of 25 and 5, respectively. As a major advantage of the two-stage integrator circuit, chopping can be used for offset and noise reduction. In contrast to conventional chopping, the chopper frequency can be lower than the signal bandwidth. Experimental results confirm that chopping frequencies > 5 MHz disappear in the noise floor and an output noise voltage of 2.2 mVrms is reached. The implemented Rogowski coils with the two-stage integrator circuit cover the required bandwidth for on-chip and off-chip current sensing and are suitable for real-time current sensing. This was confirmed by transient measurements with sinusoidal signal current up to 70 MHz as well as for current pulses of 60 A at a slope of 1 kA/μs.

Based on the implemented open-loop sensing concept, this work introduces the first known compensated sensor front-end for a Rogowski coil. Thus, a significantly higher sensitivity of 447 mV/A, a larger linear sensing range and a smaller output offset of <1.5 mV with a comparable bandwidth from 17 kHz to 63 MHz is achieved. The output signal of the Rogowski coil (sensor front-end input) is compensated by a feedback loop in the form of a high-pass filter. Compensation up to high frequencies is achieved by a pseudo-differential amplifier with a bandwidth of >250 MHz. This allows to extend the linear sensing range significantly at high frequencies. Furthermore, the sensing circuit output offset is improved by an additional low-pass feedback by a factor of 69 and reaches $<1.5\,$mV compared to open-loop current sensing with $104\,$mV. For noise reduction of about 93 %, auto-zeroing and copping are used for the different signal paths.

Hall sensors are utilized to extend the limited low-frequency range of Rogowski coils towards DC signal currents. As an advantage of a Hall sensors their sensitivity can be adjusted to the sensitivity of the Rogowski coil by the bias current.

In this work, dedicated IC level Hall devices have been designed for on-chip and off-chip current sensing in the given $180\,$nm HV CMOS technology. For on-chip current sensing, a cross-shaped Hall device is placed besides a power line to measure internal signal currents. For TCAD simulations of the Hall device, parameters of the particular technology are extracted from measurement results. The sensitivity of approximately $10\,$mV$/$T at a bias current of $1\,$mA was confirmed with both TCAD simulations and experimental results. Due to the geometry of the power line and the distance to the Hall device, a current-related sensitivity of $24\,\mu$V$/$A was measured for on-chip current sensing. For off-chip current sensing, in this work, a vertical Hall device is implemented in a $60\,\mu$m back-grinded chip. Experimental results confirm a current-related sensitivity of $8.3\,\mu$V$/$A of the Hall device, placed on top of a power line.

A capacitively coupled sensor front-end is introduced. Thus, the same amplification can be guaranteed by the front-end for different Hall devices and different bias currents. Measurements confirm that, by modulating the Hall voltage into the AC range and low-pass filtering, the offset and noise of the front-end is reduced. In order to cover a wide frequency range (kHz–MHz) for AC modulation, a high-pass filter with a pseudo resistance of $7.5\,$GΩ is implemented. The symmetrical layout of the planar cross-shaped Hall device for on-chip current sensing allows four-phase current spinning for signal modulation and offset reduction of the Hall device. A maximum modulation frequency of $10\,$MHz is ensured by controlling the current spinning by a non-overlapping clock with a $5\,$ns delay. While for the vertical Hall device, for off-chip current sensing, chopping is implemented for AC modulation. Experimental results confirm that the limited low-frequency range of the Rogowski coil for current sensing can be extended by the Hall sensors towards DC. The on-chip current sensing covers a frequency range from DC to $27\,$kHz and the off-chip current sensing covers a frequency range from DC to $43\,$kHz. The sensitivity of $43.3\,$mV$/$A for on-chip current sensing is achieved by a bias current of $3.3\,$mA of the Hall device, whereas the vertical Hall device requires a bias current of $16\,$mA to achieve the sensitivity of $3.1\,$mV$/$A.

Finally, a hybrid current sensor consisting of the previously introduced Hall path and Rogowski path is introduced, connected by a high-pass filter. This allows wide-bandwidth current sensing from DC up to $71\,$MHz with a sensitivity of $29.5\,$mV$/$A, a low output noise of $1.4\,$mVrms. The measured bandwidth is slightly below the expected bandwidth of $75\,$MHz due to the capacitive load from the interconnection network, whereby the output noise is improved by this interconnection network. The bandwidth of $71\,$MHz exceeds the bandwidth of fully integrated state-of-the-art hybrid current sensors by factor 23 and of discrete state-of-the-art hybrid current sensors by factor 1.4.

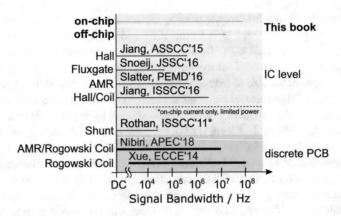

Fig. 8.1 Bandwidth comparison of this book with state-of-the-ate current sensor principles

In order to finally compare this work with prior art based on existing the sensor principles, shown in Sect. 2.2 (Fig. 2.10), Fig. 8.1 is extended by the achieved bandwidth of the on-chip current sensing and off-chip current sensing. This work constitutes the fastest current sensor on IC level and achieves a comparable bandwidth to PCB level implementations.

Index

A

Amplifier
  capacitively coupled, 117, 120
  pseudo-differential symmetrical, 84, 94
  symmetrical, 66, 80, 86, 95, 121, 123
  symmetrical cascoded, 67
Auto-zeroing, 2, 70, 75, 87
  ping-pong, 72, 74, 88

B

Back-grinded chip, 48, 50, 75, 95, 112

C

Capacitively transmission, 121, 125
Carrier mobility, 107, 109
Chopping, 2, 65, 72, 74–76, 87, 94, 118, 121
  signal modulation, 72
Clock generation, 118
Common-mode feedback (CMFB), 67, 85, 95, 124
Common-mode feedforward (CMFF), 85
Common-mode rejection ratio (CMRR), 64, 85
Connection network, 125
Current spinning, 117, 118, 121

D

Depletion region, 105
Dopant density, 102–104, 107

E

Electrical charge, 102
Electromagnetic interference, 10

F

Faraday's law, 25
Fault detection, 10
Ferromagnetic material, 13, 14
Fluxgate, 14, 95

G

Gallium nitride (GaN), 1, 10
Geometric mean distance (GMD), 30, 35, 42, 49, 52, 109, 113

H

Hall
  operation principle, 101
Hall coefficient, 102
Hall current sensor, 95, 101
  calibration, 110
  trimming, 109
Hall device
  bias current, 101, 106, 111, 117, 118, 121
  cross-shaped, 103, 105, 107, 117, 125
  geometry, 102, 103, 105
  geometry factor, 103, 105, 114
  off-chip current sensing, 111–113
  offset voltage, 103
  on-chip current sensing, 103–111
  resistance, 105, 107, 117
  sensitivity, 103, 105–109, 111, 113, 121
  temperature drift, 109, 112
  vertical, 111, 118
Hall effect, 15, 101
Hall factor, 102, 107–109

© The Editor(s) (if applicable) and The Author(s), under exclusive license to
Springer Nature Switzerland AG 2020
T. Funk, B. Wicht, *Integrated Wide-Bandwidth Current Sensing*,
https://doi.org/10.1007/978-3-030-53250-5

Hall path
  bandwidth, 122
  front-end, 120–122
  off-chip current sensing, 118
  on-chip current sensing, 117–118
  sensitivity, 122
  transfer behavior, 122
Hall voltage, 102, 103, 107, 109, 111, 117, 121
Helmholtz coil, 106
High-frequency structure simulator (HFSS),
    42, 45
Hybrid current sensor, 17, 125–127
  bandwidth, 126
  on-chip sensing, 125
  sensitivity, 126
  transfer behavior, 125

I
Insulated-gate bipolar transistor (IGBT), 1, 9,
    77

J
Junction capacitance, 102

L
Lorenz force, 15, 101

M
Magnetic field, 42, 48, 101, 103, 106, 109, 111,
    118
Magnetoresistive resistor (MR), 13
Multi-dimensional magnetic field
    measurement, 112

N
Noise, 69, 90, 94, 103, 117, 118, 127
  $1/f$ noise, 2, 70, 73, 87, 121
  floor, 73
  power spectral density (PSD), 70
  thermal, 70, 88, 121
N-well, 15, 102, 104, 107, 111

O
Off-chip current sensor, 11, 17, 101, 121
Offset, 2, 69, 86, 90, 117, 118
On-chip current sensor, 11, 17, 101, 108, 121

P
Phase shift, 63, 81
Power electronic applications, 1, 7–10, 12
Power stage, 8
Pseudo-resistor, 121

R
RF broadband power amplifier, 75, 126
Rogowski coil
  bandwidth, 16, 27, 33, 36, 39, 51, 54, 55,
      57
  damping coefficient, 27
  damping resistor, 26, 33, 38, 51, 55, 64
  design trade-off, 29, 38–42, 48, 51, 54–59
  equivalent circuit, 25, 44, 54, 59, 82
  geometry model, 30
  helix-shaped coil, 48–59, 77
  layout, 44, 45, 49, 58
  mutual inductance, 25, 28–32, 49–52
  off-chip current sensor, 48–59
  on-chip current sensor, 28–45
  operation principle, 16, 25
  output voltage, 78
  parasitic components, 33–38, 44, 51–55
  parasitic coupling, 44, 64
  planar coil, 28–45, 76, 78, 81, 90, 125
  sensitivity, 28, 33, 39, 51, 55
  3D field simulation, 42
  transfer behavior, 16, 26, 39, 42, 45, 55, 57,
      78
Rogowski path
  AC signal compensation, 84–86
  amplitude sensing range, 76, 79
  bandwidth, 86, 94
  closed-loop sensing, 78, 79, 90
  DC gain, 63
  DC signal compensation, 86, 90
  feedback, 79, 81
  forward path, 80
  front-end concepts, 63
  high-pass feedback, 84, 94
  integrating front-end, 16, 63, 65, 76
  integrator, 66, 67, 79
  noise cancelation, 87–88
  off-chip sensing, 69, 75, 77
  on-chip sensing, 68, 75, 90
  open-loop sensing, 65, 77, 90, 94
  overcompensation, 66, 67
  sensitivity, 76, 90, 94
  signal compensation, 79, 90
  spectrum, 73

stability, 82, 95
step response, 76
transfer behavior, 63, 65, 66, 68, 86, 90
transfer function, 79
two-stage integator, 68

S
Sense-FET, 8, 12, 95
Shift register, 118
Shunt, 8, 11, 95
   compensation network, 11
Signal processing, 63, 117

Silicon carbide (SiC), 1, 10
Smart sensor, 2
State-of-the-art current sensors, 17, 94, 95
State-of-the-art hybrid current sensors, 127

T
Technology computer-aided design (TCAD)
   simulation, 103–107

W
Wheatstone bridge, 13